INSTITUTE OF CERAMICS

TEXTBOOK SERIES

1. RAW MATERIALS

INSTITUTE OF CERAMICS

TEXTBOOK SERIES

1. RAW MATERIALS

by

W. E. Worrall

M.Sc., Ph.D., A.R.I.C., Senior Lecturer in Ceramics, The Houldsworth School of Applied Science, The University of Leeds

MACLAREN AND SONS LTD.,
LONDON, ENGLAND

85334 191 5
RAW MATERIALS
(INSTITUTE OF CERAMICS
TEXTBOOK SERIES)
BY W. E. WORRALL
THIS 2ND EDITION PUBLISHED 1969
FIRST PUBLISHED 1964 BY
MACLAREN AND SONS
7 GRAPE STREET
LONDON WC2

PRINTED LITHOGRAPHICALLY IN GREAT BRITAIN
BY HOLLEN STREET PRESS LTD,
SLOUGH, BUCKS.

General Foreword

It is a recognised responsibility of a professional qualifying Institute not only to elect to membership persons of suitable qualification, but also to assist potential members to obtain the necessary qualifications. The Institute of Ceramics, having set up examinations for its Licentiate and Associate grades, and having prescribed a suitable syllabus, found itself in some difficulty owing to the lack of suitable textbooks.

In order to fill the gap, it was decided to commission a number of authors, each expert in his own topic, to prepare a series of books dealing with some of the more important aspects of ceramics. The topics chosen were Raw Materials, Rheology, Drying, Action of Heat, Glazes, and Properties of Fired Materials.

Although each book in the series stands as a separate entity in its own right, there has been sufficient collaboration between the authors to ensure that there is a minimum of overlapping and that there will be a reasonable parity in treatment throughout the series.

The primary aim is to cover the needs of the Institute's examination system, but it is clear to me that in the event a much wider purpose will have been served. The books should prove to be of great value to all ceramic technologists, whether working in the laboratory, in the factory, or in the classroom. I strongly commend them to all workers in ceramics. They have a very important contribution to make to the up-to-date, formal and systematic presentation of current knowledge in this important subject.

A. DINSDALE
President
Institute of Ceramics.

TITLES IN THE SERIES

RAW MATERIALS

by W. E. Worrall, *M.Sc., Ph.D., A.R.I.C., Senior Lecturer in Ceramics, The Houldsworth School of Applied Science, The University of Leeds.*

THE EFFECT OF HEAT ON CERAMICS

by W. F. Ford, *Ph.D., B.Sc., F.I. Ceram., Senior Lecturer in Refractories Technology, University of Sheffield.*

RHEOLOGY OF CERAMIC SYSTEMS

by F. Moore, *M.Sc., Senior Scientific Officer, British Ceramic Research Association.*

DRYING

by R. W. Ford, *B.Sc., Senior Scientific Officer, British Ceramic Research Association.*

PROPERTIES OF FIRED MATERIALS

by A. W. Norris, *B.Sc., F.Inst.P., F.I. Ceram., Managing Director,* Doulton Research Ltd.

GLAZES

In preparation. Author to be announced.

LIST OF CONTENTS

1. Raw Materials

1.1 INTRODUCTION

The oldest ceramic raw material is undoubtedly clay. Clay has been defined as an earth that forms a coherent sticky mass when mixed with water; when wet, this mass is readily mouldable but if dried it becomes hard and brittle and retains its shape. Moreover, if heated to redness, it becomes still harder and is no longer susceptible to the action of water. Such a material clearly lends itself to the making of articles of all shapes.

Clay may take various forms; it is easily recognized as the sticky, tenacious constituent of soil, but it can also occur as rock or slate which, owing to compression, is so hard and compacted that water penetrates it very slowly. Clay is seldom pure, but the substance responsible for its characteristic properties is usually called the "clay mineral" or "clay substance".

Although probably the earliest ceramic articles were made entirely from clay, from very early times additions to it of other materials are known to have been made. At the present time in the pottery industry the chief raw materials used in conjunction with clay are the various fluxes and silica. In the refractories industry, increasing demand for specialized refractories has resulted in products containing little or no clay, such as alumina, magnesite and chrome; such products are also classed as "ceramics" because in general they are shaped whilst wet and then fired in order to be hardened.

In recent years a wide variety of inorganic, non-metallic materials has been developed for the electrical, nuclear power, and engineering industries. In the shaping and processing of these products some form of heat treatment is involved, and they too are regarded as ceramic materials. Examples are: rutile, a form of titanium dioxide used for making ferroelectric materials; steatite or talc, for electrical insulators; alumina, zirconia, thoria and beryllia as refractories and electrical insulators, uranium oxide as a nuclear-fuel element, and nitrides and carbides as abrasives or insulators.

1.2 SILICA

1.2.1 Structure of Crystals

Before being able to understand the properties of silica, we must know something of its structure—that is, the way in which the solid is built up from the atoms of silicon and oxygen. The principles outlined here will help in understanding the structures of the other minerals to be described later.

Most solid substances are crystalline—their component atoms are arranged in a regular array which is called a *lattice*. A lattice is a continuous structure, capable of indefinite extension in at least two dimensions, rather like the pattern on a wallpaper that is repeated again and again. This "repeat pattern" is called a unit cell. Even in a simple lattice such as that of NaCl, however, there are no individual molecules of NaCl as such; no single atom can be said to "belong" entirely to any other atom; it is "shared" between all its neighbours.

In most types of crystal the atoms, by the loss or gain of an electron, have become charged and are therefore to be regarded as *ions*. In sodium chloride, for instance, the ions are Na^+ and Cl^-, and the electrostatic attraction of the oppositely-charged ions forms a bond, holding the crystal together; such a bond is said to be *electrovalent*. The bonding in silica and silicates is mainly of this type, but in some crystals (diamond, silicon carbide, zinc sulphide) the atoms are bonded, not by electrostatic forces, but by "electron sharing" between neighbouring atoms; this bonding is said to be of the *covalent* type and is common in organic compounds.

The sign of the electrostatic charge depends on the nature of the atom—in general, "metallic" ions carry a positive charge and non-metallic ions carry a negative charge, equal in magnitude to their valencies ; thus in silica the ions are Si^{4+} and O^{2-}. Since the lattice as a whole must be electrically neutral, the sum of the positive charges must equal the sum of the negative charges.

In ionic crystals such as sodium chloride and silica, the various types of ion may be regarded as spheres having definite radii; some examples of ionic radii are given in Table 1.

It will be clear from this table that the negative ions (anions) are in general larger than the positive ions (cations). Consequently the type of lattice largely depends on the packing of the anions.

Table 1. Ionic Radii of some Common Elements

Element	*Symbol of ion*	*Valency*	*Ionic Radius (Å)**	*Usual co-ordination number(s)*
Boron	B^{3+}	3	0.20	3, (4)
Beryllium	Be^{2+}	2	0.31	4
Silicon	Si^{4+}	4	0.41	4
Aluminium	Al^{3+}	3	0.50	4, (5), 6
Magnesium	Mg^{2+}	2	0.65	6
Sodium	Na^{+}	1	0.95	6, (8)
Titanium	Ti^{4+}	4	0.68	4, 6
Zirconium	Zr^{4+}	4	0.80	6, (8)
Calcium	Ca^{2+}	2	0.99	(7), 8, (9)
Potassium	K^{+}	1	1.33	(6), (7), 8, (9), (10), 12
Iron: ferrous	Fe^{2+}	2	0.75	6
ferric	Fe^{3+}	3	0.60	4, 6
Oxygen	O^{2-}	—2	1.40	—
Fluorine	F^{-}	—1	1.36	—
Sulphur	S^{2-}	—2	1.84	—
Chlorine	Cl^{-}	—1	1.81	—

* Å = 1 Ångstrom Unit = 10^{-10} m. ( ) signifies less common co-ordination numbers.

Another important feature of the lattice is that, for maximum stability, each ion tends to surround itself with the greatest possible number of oppositely-charged ions, so that all are in contact; this number is known as its co-ordination number. The maximum number of neighbouring ions are in contact. Since Si^{4+} is a small ion, the maximum number of O^{2-} ions that can surround it, all being in contact, is four. We say therefore that the *co-ordination number* of silicon with respect to oxygen is four. This arrangement is shown in perspective in Figure 1.

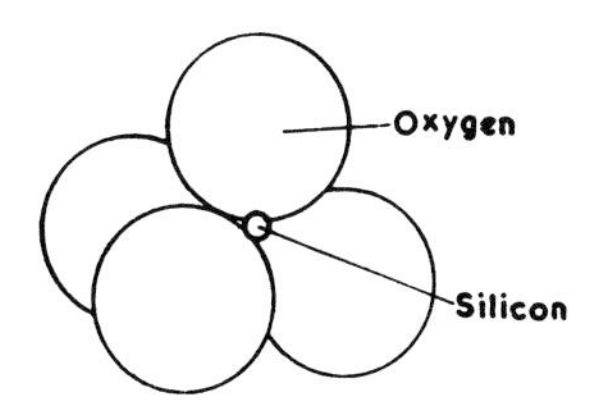

Fig. 1.—Packing of silicon and oxygen atoms in silica and the silicates.

Each type of ion, depending on its size, has a characteristic co-ordination number with respect to oxygen; co-ordination numbers for some of the commoner cations with respect to oxygen are given in Table 1.

The arrangement of four O^{2-} ions around one Si^{4+} ion is characteristic of all compounds containing silicon and oxygen. If we imagine the centres of the four oxygen ions to be joined by straight lines, as in Figure 2, the resulting geometrical figure is known as a tetrahedron, with a triangular base and three triangular sides meeting at an apex; the Si^{4+} ion is situated at the centre of this tetrahedron. Often it is convenient to speak of this (SiO_4) unit as a "silica tetrahedron".

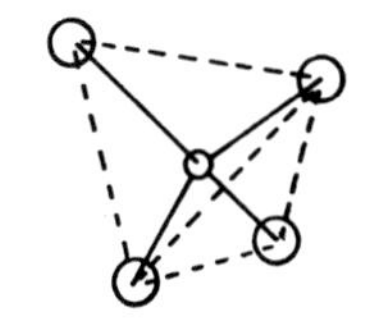

Fig. 2.—A silicon–oxygen tetrahedron.

We can work out the "valency requirements" as follows. Each Si^{4+} ion, of course, has four "valency bonds", and may be imagined to "share" these four valency bonds between the four surrounding O^{2-} ions, giving one "valency share" (+1) to each. (Figure 3).

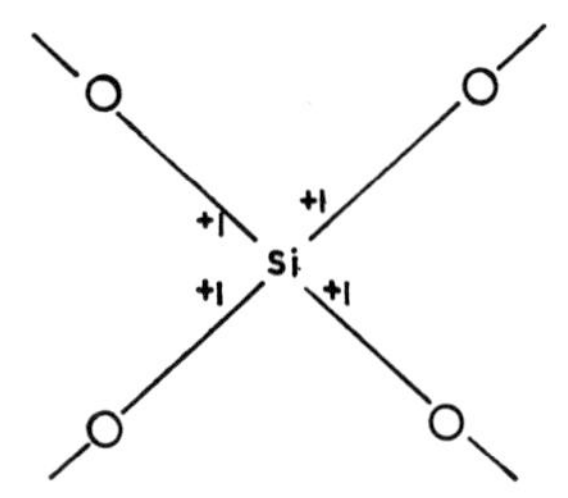

Fig. 3.—Distribution of charges in the silicon–oxygen tetrahedron.

In this way the valency requirements of Si^{4+} are satisfied, but the oxygen valencies in this (SiO_4) group are only partially satisfied, because whereas each O^{2-} requires two valency shares, it receives only one. In order therefore to satisfy the free oxygen valencies, the (SiO_4) group requires other ions. In silica (SiO_2) these free valencies are satisfied because the tetrahedra join to one another in such a way that each O^{2-} ion is effectively joined to two Si^{4+} ions. Consequently the valency requirements of both Si^{4+} and O^{2-} are satisfied, and the net formula is then not (SiO_4) but SiO_2. There are several ways in which the tetrahedra can be joined, and it is therefore not surprising that there are three principal crystalline forms of silica—quartz, cristobalite and tridymite—each having the formula SiO_2, but differing in the way the silica tetrahedra are arranged.

1.2.2 Structure of Quartz

In quartz the Si–O–Si bonds joining neighbouring tetrahedra are not straight but bent round to form spiral chains (Figure 4). Starting with any Si^{4+} ion, a spiral can be traced in one definite direction, depending on the crystal (in this instance the direction is anti-clockwise). The whole structure consists of many such

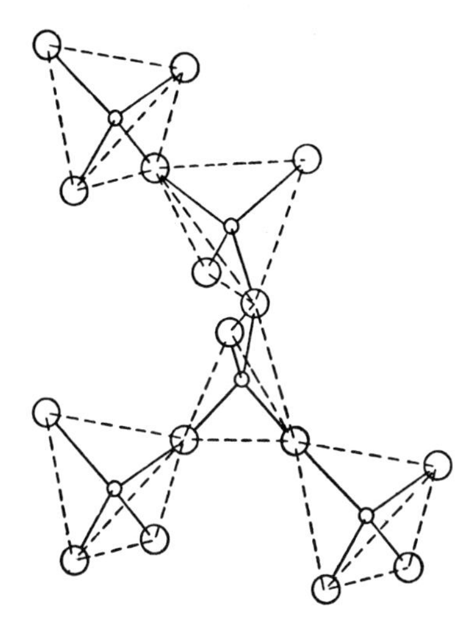

Fig. 4.—Arrangement of silica tetrahedra in quartz.

spiral chains, joined by common Si^{4+} ions, shown in perspective in Figure 5. Note that, in these and subsequent atomic models, the distances between atoms have been exaggerated and the ionic radii are not to scale. This has been done to show the valency bonds clearly; in a scale model with all the atoms in contact the bonds would be difficult to discern.

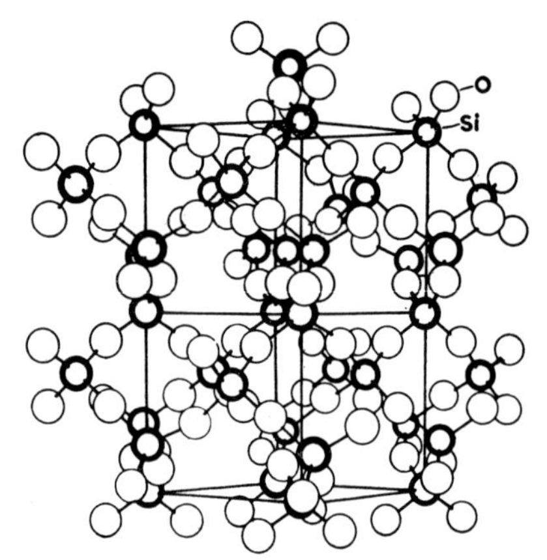

Fig. 5.—The structure of β-quartz. (Atoms nearer to the reader drawn in heavy outline).

1.2.3 Structures of Tridymite and Cristobalite

The structures of tridymite and cristobalite are perhaps easier to understand. In both these structures the silica tetrahedra form rings containing six Si atoms and six O atoms, as shown in Figure 6, A and B. What the figure cannot show is that in neither

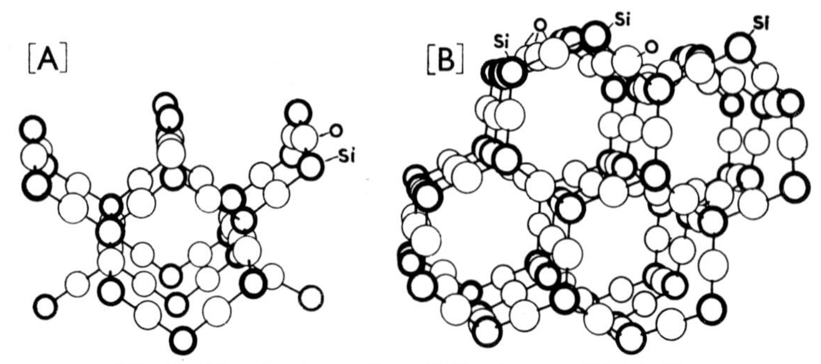

Fig. 6.—The structures of : A. Tridymite. B. Cristobalite.

the tridymite nor the cristobalite structure are the rings flat; they are somewhat distorted and so the Si atoms are not all in the same plane. This distortion is more marked in cristobalite than in tridymite. The difference between the two is best appreciated by considering how the oxygen atoms are arranged. Figure 7A shows how two silica tetrahedra are linked in tridymite. Note that the lowest three oxygen atoms form the triangular base of the lower tetrahedron. The second tetrahedron is inverted, with its base uppermost, and is joined to the first through a common oxygen atom forming the apex. Now observe that the three basal oxygen atoms of the upper tetrahedron fall directly below corresponding oxygen atoms in the base of the lower tetrahedron; this arrangement of the oxygen repeats throughout the structure and is characteristic of tridymite.

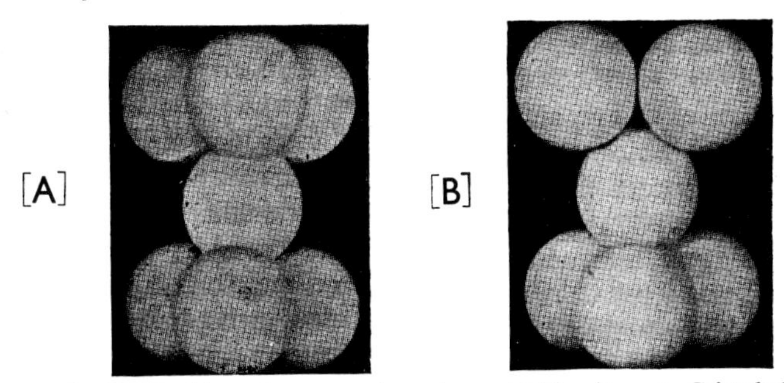

Fig. 7.—Packing of oxygen atoms in : A. Tridymite. B. Cristobalite.

Figure 7B shows the arrangement of the oxygen atoms in cristobalite and it is clear that the basal oxygen atoms of the respective tetrahedra no longer correspond; it is as if one of the tetrahedra of Figure 7A had been twisted through 60° with respect to the other. This arrangement is characteristic of the oxygen atoms in cristobalite.

1.2.4 Conversions

The structures of the three forms of silica differ considerably, and it is not easy to convert one to another. Any such conversion involves the breaking of Si–O– bonds—in other words, separating the tetrahedra and then joining them again differently. Nevertheless, if quartz is heated above 1470°C for a considerable time, it is gradually converted to cristobalite. Furthermore, if cristobalite is heated in the range 870°C to 1470°C, it is gradually converted to tridymite. Both these reactions can be speeded up by catalysts or mineralizers —e.g. lime assists the conversion of quartz (or tridymite) to cristobalite. These conversions may be represented by the equation:

$$\text{quartz} \xrightarrow{1470^\circ} \text{cristobalite} \overset{870^\circ}{\rightleftharpoons} \text{tridymite}$$

these high-temperature forms of silica can only with great difficulty be reconverted to quartz, consequently in the equation the conversion of quartz to tridymite is represented by a single arrow. This conversion being apparently irreversible, it is surprising that quartz is much more common in Nature than the other forms; cristobalite and tridymite are, of course, formed artificially, e.g. in the manufacture of silica bricks. All three forms are relatively stable at room temperature; only at 870°C and above are the Si–O bonds sufficiently "loose" for conversions to take place.

1.2.5 Inversions

Although the three forms of silica are relatively stable within the temperature ranges mentioned, certain minor changes in structure do occur, which are of sufficient practical importance to deserve attention. For instance, if quartz is heated to 573°C or over, the Si–O–Si bonds straighten somewhat (Figure 8). As a

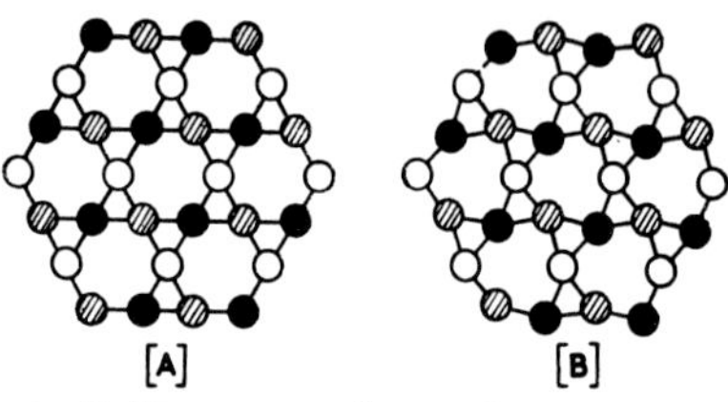

Fig. 8.—Arrangement of silicon atoms in : A. β-quartz. B. α-quartz. (Reproduced from "Structural Inorganic Chemistry", by kind permission of A. F. Wells and the Clarendon Press, Oxford).

result, the atoms become less closely packed and a marked expansion occurs. The modification that exists at room temperature is known as α-quartz and that above 573°C is known as β-quartz, and the transition from one to the other can be represented by the equation:

$$\alpha\text{-Quartz} \xrightleftharpoons{573°C} \beta\text{-Quartz.}$$

By an exactly similar process, both tridymite and cristobalite undergo inversions as follows:

$$\alpha\text{-Cristobalite} \xrightleftharpoons{220°\text{–}260°C} \beta\text{-Cristobalite}$$

$$\alpha\text{-Tridymite} \xrightleftharpoons{117°C} \beta_1\text{-Tridymite} \xrightleftharpoons{163°C} \beta_2\text{-Tridymite.}$$

All these inversions are reversible, as the equations imply, and occur very rapidly, in contrast to the conversion of one principal form of silica to another.

1.2.6 Other Forms of Silica

Although in recent years, certain new high-pressure forms of silica have been discovered, it is sufficient here to mention only the following more common forms:—

Silica Gel

When solutions of sodium silicate are acidified, silica is precipitated as a gelatinous mass, which can then be extracted and dried. The resulting silica gel, as it is called, has no definite structure and is non-crystalline. The structural units of silica gel are still silica tetrahedra but they are joined together at random and not according to a "pattern".

Vitreous Silica

When quartz, cristobalite or tridymite is heated above 1710°C (the fusion point) and then cooled rapidly, again the silica tetrahedra do not have time to arrange themselves in a definite order, and so they link in a random fashion to form a glass (more accurately a "super-cooled" liquid) which, like silica gel, is also amorphous. It is therefore not strictly correct to call it "fused quartz", as is commonly done.

1.2.7 Physical Properties

Pure quartz occurs as transparent, hexagonal crystals, and at 20°C has a specific gravity of 2.65.

Under the microscope, in thin sections of fired siliceous material tridymite usually appears as wedge-shaped crystals; occasionally it has been found in volcanic lava. In tridymite the atoms are packed less densely than in quartz, and therefore the specific gravity of tridymite is lower—2.27 at 20°C.

Cristobalite, like tridymite, is rarely found in Nature, but under the microscope in thin sections of certain ceramics it is frequently observed as a mass of small crystals. In cristobalite the packing of the atoms is also less dense than in quartz, the specific gravity of cristobalite at 20°C being 2.33.

The variation in the coefficient of thermal expansion of quartz and cristobalite with temperature reveals very strikingly the α–β inversions, and is best seen by plotting the percentage volume expansion for each mineral against the temperature (Figure 9). Owing to the difficulty of obtaining pure tridymite its expansion curve has been omitted.

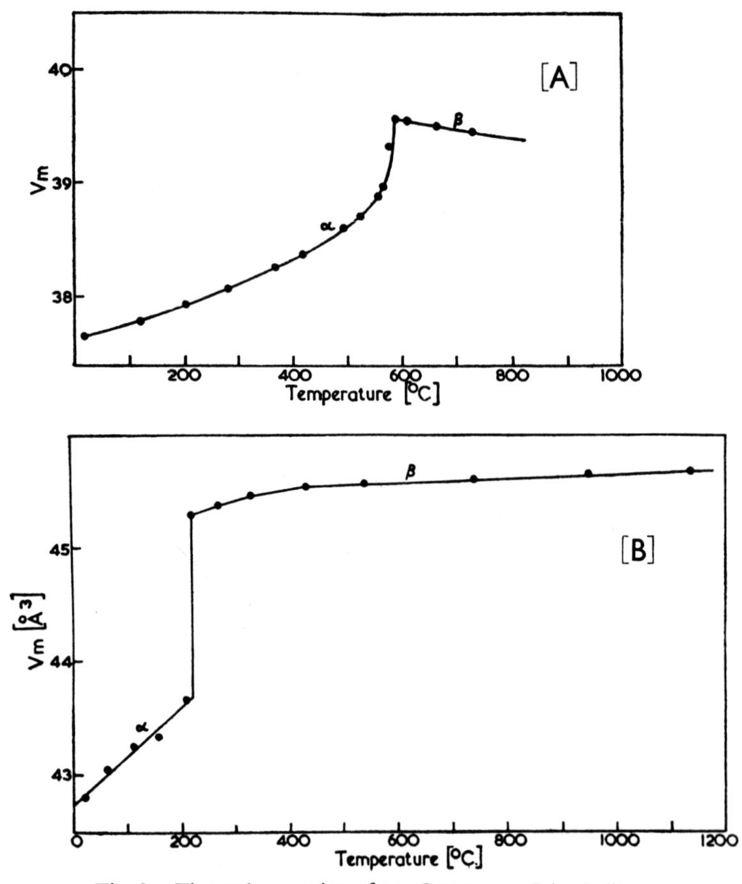

Fig. 9.—Thermal expansion of : A. Quartz B. Cristobalite.

At the inversion temperatures there is clearly a marked increase in the rate of expansion for each mineral, particularly for the α–β quartz change at 573°C and for the α–β cristobalite change at 220°–260°C. These sudden expansions are the cause of spalling in silica refractories.

Since the linear expansion of silica at 1000°C is about twice that of clay, the expansion of a pottery body is often purposely increased by the addition of flint, which on being fired transforms to cristobalite. Of all the forms of silica referred to, vitreous silica expands the least (0.05% linear between 20° and 1000°C) and on this account it withstands sudden changes of temperature without shattering. Vitreous silica is therefore useful for making laboratory ware such as crucibles and tubes.

1.2.8 Chemical Properties

Crystalline silica resists the action of most aqueous alkalis and acids, apart from hydrofluoric acid, which reacts with silica to form silicon tetrafluoride, SiF_4, which is a volatile substance:

$$4HF+SiO_2=\uparrow SiF_4+2H_2O$$

This reaction is made use of in the gravimetric determination of silica. However, crystalline silica is attacked by fused caustic alkalis, and alkaline slags, forming silicates; for example, when silica is fused with sodium hydroxide, sodium silicate is formed. Fortunately crystalline silica is not readily attacked by ferric oxide, hence silica bricks are used in open-hearth steel furnaces.

Whereas the reactivity of silica glass with chemical reagents is similar to that of crystalline silica, silica gel is much more reactive, presumably because it is more finely divided. Apart from its chemical reactivity silica gel is a powerful water-absorbant (hence it is used as a desiccant) and is attacked by aqueous alkalis.

1.2.9 Occurrence

Silica occurs in nature: as *quartzite rock* and *ganister*, both being composed principally of quartz; as *sand* or *sandstone*, also consisting of quartz; as *flint pebble*, which is a crypto-crystalline quartz (i.e. the crystals are too small to be seen under the optical microscope but can be detected in other ways); and as chalcedony, opal and agate, which are also crypto-crystalline quartz.

Quartzite rock is found in North and South Wales, the Northern Pennines and elsewhere; it belongs to the Millstone Grit Formation.

Ganister, forming the seat-earth of coal seams, is found in the Sheffield district. Like quartzite, ganister is a Carboniferous deposit, situated underneath the Coal Measures, and is found chiefly in Derbyshire, Yorkshire, and Gartcosh (Scotland). It consists of very small particles of quartz, contaminated with a little clayey material, which provides just sufficient plasticity to bind the ground mass together when moistened, and it is used for making silica bricks. Both quartz and ganister contain 97% or more of silica.

Flint is an important form of silica, used in the pottery and refractories industries. It is believed to consist of small crystals of quartz, bound together by molecules of water and their presence probably accounts for the specific gravity being about 2.62, slightly less than that of quartz.

When flint is calcined, the combined water is driven off about 400°C, causing the large aggregates to lose their strength and become crumbly; finally, at 1100°C, the quartz changes to cristobalite, this conversion being promoted by the presence of calcium oxide, an impurity.

Flint occurs as nodules in the Upper and Middle Chalk and as pebbles on beaches adjacent to the Chalk; they are largely of Cretaceous origin (see Table 9).

1.3 CLAYS

1.3.1 The Clay Minerals

In the introduction it was pointed out that clays are seldom pure; in addition to the clay mineral, which is the essential and preponderant substance, and so is responsible for the characteristic properties, a number of adventitious minerals such as quartz, mica, and iron oxide are present.

In order to gain a clear idea of the clays from which ceramics are manufactured, it is necessary to study the structures and properties of their clay minerals, but it must be remembered that the properties of any clay in the bulk will also depend on the nature and proportion of the impurities.

1.3.2 Structure of the Clay Minerals

There are two main groups of clay mineral: the *kaolins* and the *montmorillonites*. The kaolins have the empirical formula $Al_2Si_2O_5(OH)_4$, and that of the montmorillonites can be derived from the formula $Al_2Si_4O_{10}(OH)_2$, as described later.

The kaolin group includes kaolinite, nacrite, dickite and halloysite; halloysite can also exist in a hydrated form having the formula $Al_2Si_2O_5(OH)_4.2H_2O$. Of all these, by far the most important is kaolinite, since it is the principal constituent of china clay, ball clay, fireclays, and many brick clays; the other kaolin minerals are somewhat rare in Britain.

1.3.3 The Kaolin Group

All the kaolin minerals have the same basic structural features: they consist of a layer of Si–O atoms, referred to as the silica or tetrahedral layer, joined by common O atoms to a similar layer of Al–O atoms, called the gibbsite or octahedral layer. The arrangement of the atoms in these two layers are shown in Figures 10A and 10B respectively. The silica layer closely resembles that of cristobalite and tridymite (see Figure 6). In the silica layer

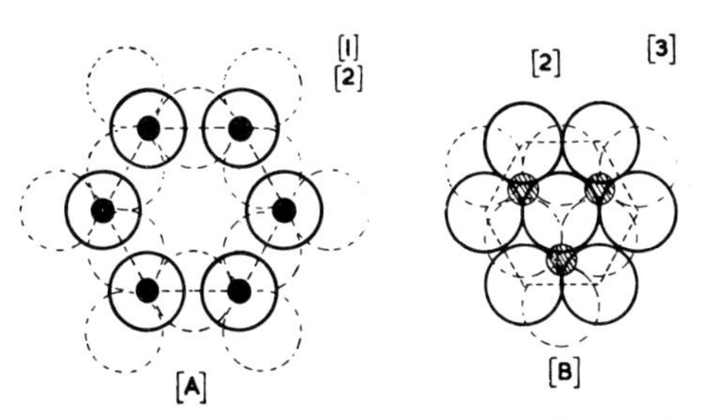

Fig. 10.—The structure of the kaolins : arrangement of atoms in : A. The silica layer. B. The gibbsite layer. (Reproduced from "Structural Inorganic Chemistry", by kind permission of A. F. Wells and the Clarendon Press, Oxford).

(Figure 10A) the Si and O atoms are linked to form six-membered rings, similar to those in cristobalite and tridymite. The Si atoms are, as usual, in fourfold co-ordination with O, and the vertices of all the silica tetrahedra point upwards.

In the gibbsite layer (Figure 10B), the Al atoms are in sixfold co-ordination with O (or OH), but the O atoms, drawn in unbroken line, form a hexagonal ring of approximately the same size as the Si–O hexagonal rings. Therefore the two layers, if superimposed, "fit" almost exactly, forming a compound layer with the O atoms, referred to as "common" atoms, forming the link between them. The composite layer is shown in perspective in Figure 11. As pointed out on page 6, the sum of the positive and negative charges must be zero, as follows:

$$Al_2^{3+} Si_2^{4+} O_5^{2-} (OH)_4^{-}$$
$$6 + 8 - 10 - 4 = 0$$

The average formula of the Si–O rings is Si_2O_3 and that of the gibbsite layer is $Al_2(OH)_4O_2$. If these rings are joined with the two O atoms forming a "bridge", we arrive at the composite

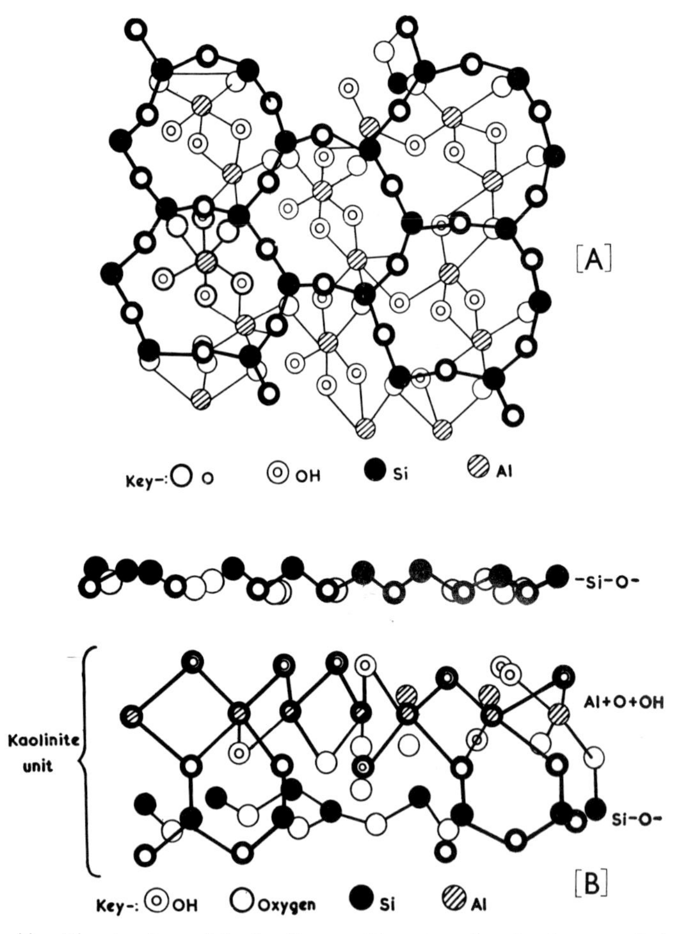

Fig. 11.—The structure of the kaolins : A. Plan view, showing hexagonal rings. B. Front elevation.

formula $Al_2Si_2O_5(OH)_4$, which is the basic formula of all the kaolin minerals.

It is important to realize that the valencies of the oxygens of the Si_2O_3 layer (the so-called planar oxygens) are completely satisfied, since each O atom is joined to two Si atoms, and so it cannot be joined by direct chemical bonds to any other atoms—in other words, the Si_2O_3 layer has no tendency to extend itself perpendicularly. Horizontally, however, it can extend indefinitely;

precisely the same applies to the planar OH groups in the gibbsite layer.

Nevertheless the thickness of a crystal of a kaolin mineral is much greater than just two composite layers; in fact, a crystal will consist of many thousands of such layers. Since, as we have just seen, there are no unsatisfied valencies by means of which one kaolin unit can be linked with the next, bonding must be by other means. In the kaolins the successive units are linked by so-called hydroxyl bonds, where a H atom of an OH group acts as a bridge between two adjacent O atoms. The H atom may be imagined to alternate rapidly between situations A and B, Figure 12. This bonding takes place between OH groups of the gibbsite layer and

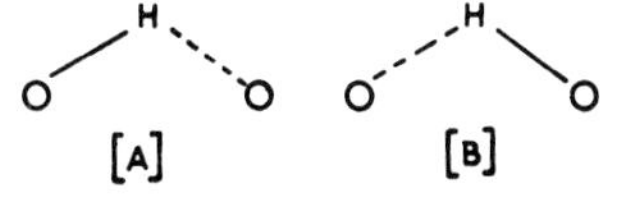

Fig. 12.—Hydroxyl bonding.

planar O atoms of the silica layer in the adjacent kaolin unit. The arrangement of units in a complete crystal will be made clearer by Figure 13.

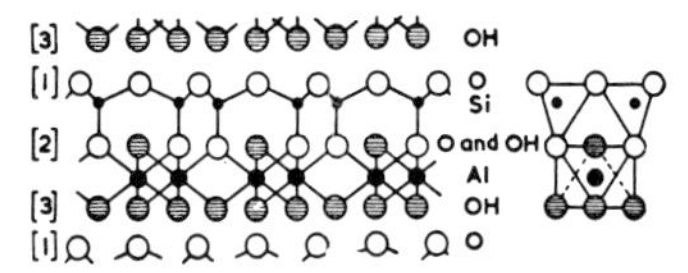

Fig. 13.—Arrangement of kaolin units in a crystal (reproduction from "Structural Inorganic Chemistry", by kind permission of A. F. Wells and the Clarendon Press, Oxford).

So far, everything that has been stated with regard to structure applies equally to all the members of the kaolin group: the distinguishing feature is the way in which the units are stacked one upon another. In Figure 13 the stacking is approximately that of the mineral nacrite, in which the O atoms in each silica layer correspond exactly, giving the most symmetrical form of crystal. A full discussion of the subject is beyond the scope of this work, and it is sufficient to state that in dickite and kaolinite the layers are displaced slightly, by a constant amount, from the "ideal" arrangement of Figure 13, whereas in halloysite the displacements are completely random. Because of this, in halloysite the bonding is so weak that water molecules can penetrate the layers, and natural halloysite is therefore normally in a hydrated form, having the formula $Al_2Si_2O_5(OH)_4.2H_2O$.

In the majority of British fireclays and ball clays, the "clay mineral" is usually kaolinite, but it is imperfectly crystalline, with some "disorder" in the stacking of the units.

1.3.4 The Montmorillonite Group

Once the student has grasped the structural features of the kaolin group, the structures of the other clay minerals and related groups will be readily understood.

Instead of one gibbsite layer being condensed with one silica layer, it is clearly possible for two silica layers to be condensed with one gibbsite layer, the latter losing two of its hydroxyl groups in the process. The formula is $Al_2Si_4O_{10}(OH)_2$, the mineral *pyrophyllite*. Now in exactly the same way it is also possible for a magnesia or brucite layer, $Mg(OH)_2$, to be condensed with two silica layers; in this case the formula is $Mg_3Si_4O_{10}(OH)_2$, and the mineral is *talc* (Figure 14). Pyrophyllite and talc can be regarded

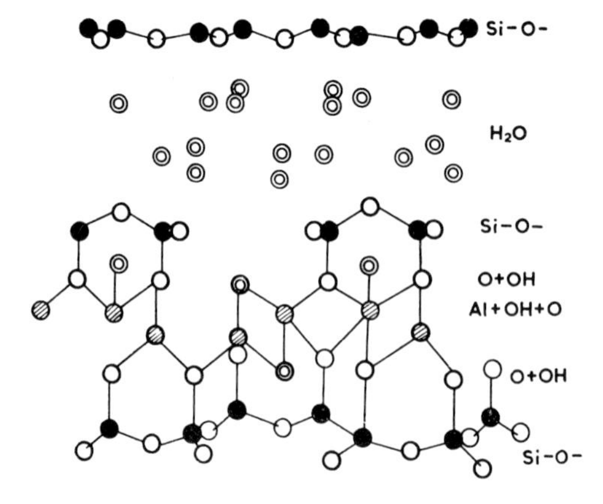

Fig. 14.—The structure of pyrophyllite and talc.

as the "parent" substances from which all the members of the montmorillonite group are derived. Now in crystal lattices generally the form or skeleton is determined by the arrangement of the largest ions—in this instance the oxygen ions. Provided that the cations are of similar size, one can replace another without this arrangement being altered; this replacement is known as isomorphous substitution. One very common example of this is the substitution of Al for Si in the feldspar structure, which is possible because Al and Si have similar radii and both can therefore "fit" into fourfold co-ordination with oxygen. Most other cations are too large for this fourfold co-ordination, but they can enter into sixfold co-ordination in place of Al—i.e. they can replace Al in the gibbsite layer: examples of such substituent ions are Mg^{2+}, Fe^{2+}, Fe^{3+} and Li^{+}. Similarly, of course, the same ions can replace Mg in the brucite layer. The degree of replacement is limited to approximately one atom in every six for each unit layer. For calculations, it is convenient to work from the unit formulae of pyrophyllite or talc, the former being usually written $Al_2Si_4O_{10}(OH)_2$. In order to show, for example, a substitution of one ion in every six by Mg, it is common practice to use fractional atoms, in order to avoid multiplying up the formula by 3. Thus this substitution would be written:

$$(Al_{1.67}Mg_{0.33})Si_4O_{10}(OH)_2,$$

which is clearly a substitution of 1 in 6.

We cannot actually perform these substitutions in the laboratory but we do find them ready made, as it were, in nature. The clay mineral montmorillonite may be regarded as having been derived from pyrophyllite, $Al_2Si_4O_{10}(OH)_2$, through replacement of some of the Al by Mg. Since, however Mg, being divalent, only carries two positive charges as against three for Al, there is a deficiency of charge which has to be satisfied by an external cation, outside the lattice. Usually it is Na, and so we may write the formula of montmorillonite:

$$Na^{+}_{0.33}[Al^{3+}_{1.67}Mg^{2+}_{0.33}] \quad Si^{4+}_4 \quad O^{2-}_{10}(OH)^{-}_2$$

$$0.33 + 5.0 + 0.67 + 16 - 20 - 2 = 0$$

This Na, being outside the lattice, can exchange with other cations (see cation exchange). Note that the part of the lattice in square brackets carries 0.33 units of negative charge, balanced by 0.33 units of positive charge, or $Na_{0.33}$. (If we wish to avoid talking about fractional atoms, we can, of course, multiply the formula throughout by 3, which, however, does not affect any of the preceding remarks). Some of the commoner members of the montmorillonite group are listed in Table 2.

Like crystals of the kaolin group, a crystal of montmorillonite is many units thick, yet clearly the unit layers cannot be bonded chemically; neither, indeed, can there be any hydroxyl bonding, since all the planar atoms on both sides of the units are oxygens. The unit layers are in fact held together by very weak "residual" forces, known as van der Waals' forces, which are caused by slight

Table 2. Empirical Formulae of some Montmorillonites

Name of Mineral	*Empirical formula*
Montmorillonite	$Na_{0\cdot33}\cdot(Al_{1\cdot67}Mg_{0\cdot33})Si_4O_{10}(OH)_2$
Nontronite	$Na_{0\cdot33}\cdot Fe(Al_{0\cdot33}Si_{3\cdot67})O_{10}(OH)_2$
Beidellite	$Na_{0.33}Al_2\ (Si_{3.67}Al_{0.33})\ O_{10}(OH)_2$
Hectorite	$Na_{0\cdot33}(Li_{0\cdot33}Mg_{2\cdot67})Si_4O_{10}(OH)_2$
Saponite	$Na_{0\cdot33}\cdot Mg_3(Si_{3\cdot67}Al_{0\cdot33})O_{10}(OH)_2$

displacements of electrical charges inside the atoms. So weak are these bonding forces that water can readily force the unit layers apart, and so when the montmorillonite clays are placed in water they swell.

The main structural features of the montmorillonites are as shown in Figure 14; they resemble those of talc and pyrophyllite, excpt that between the silica layers there are exchangeable ions and water molecules.

1.3.5 The Micas

Although the micas are not clay minerals in the strictest sense, they must be mentioned here because they so often occur in clays.

Like the montmorillonites, the micas can be regarded as derived from talc or pyrophyllite by isomorphous substitution. Imagine that we start with pyrophyllite, $Al_2Si_4O_{10}(OH)_2$, and replace one in four of the Si atoms by Al ; this substitution would result in a charged lattice having the formula $[Al_2(Si_3Al)O_{10}(OH)_2]^{-1}$. Note that the substituent Al is written separately because it is in the silica, as distinct from the gibbsite layer.

If this negative charge of one unit is balanced by the introduction of a potassium ion, K^+, the following formula results:

$$K^+.Al_2(Si_3Al)^-O_{10}(OH)_2 \text{ or } KAl_3Si_3O_{10}(OH)_2.$$

This is the formula of potash mica or muscovite, the arrangement of atoms in the structure being similar to that of pyrophyllite, except that there are K^+ ions between the silica layers (Figure 15).

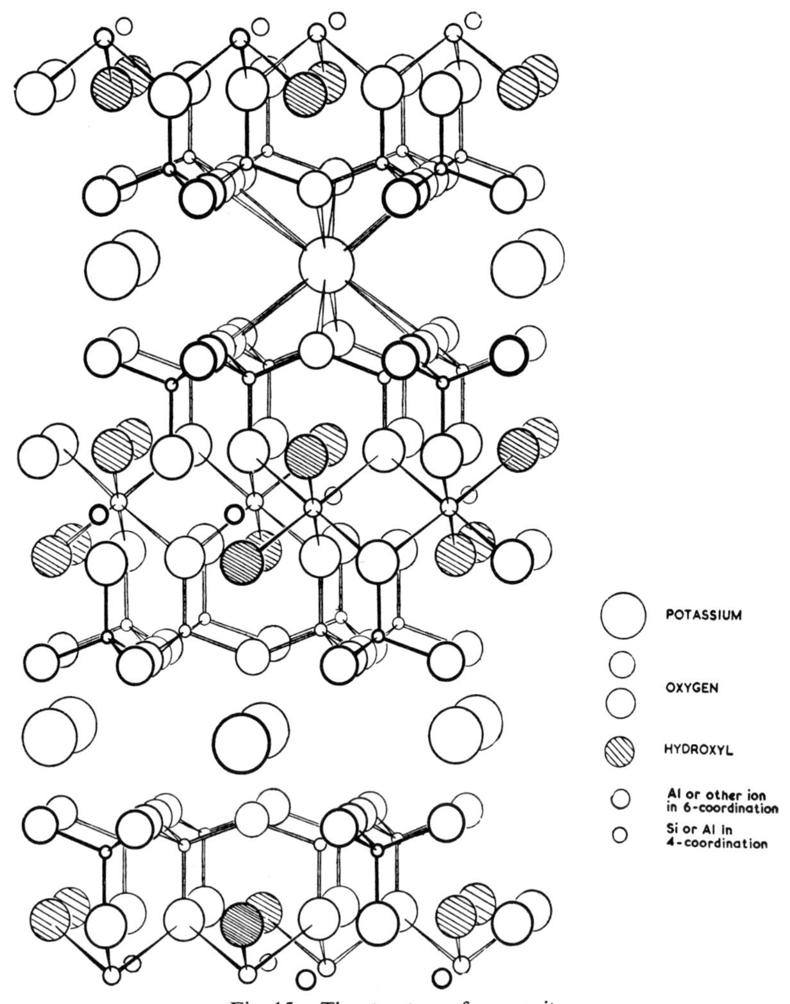

Fig. 15.—The structure of muscovite.

Following this line of reasoning, imagine a further substitution—two atoms out of four Si being replaced by Al, resulting in the charged lattice:

$$Al_2(Si_2Al_2)^{2-}O_{10}(OH)_2.$$

This formula clearly carries two units of negative charge, and must therefore be balanced by a divalent cation, for example, Ca^{2+}, giving $Ca^{2+}.Al_2(Si_2Al_2)^{2-}O_{10}(OH)_2$, which is calcium mica or margarite.

In the same way, a series of micas can be derived from talc, e.g. phlogopite, $KMg_3(Si_3Al)O_{10}(OH)_2$. Lepidolite, $K(AlLi_2)Si_4O_{10}(OH)_2$, mentioned in the section on fluxes, also falls into this group, and is an example of substitution in the gibbsite layer.

The micas do not depend on hydroxyl or van der Waals' bonding, because the positively-charged K^+ or Ca^{2+} ions hold the negatively-charged units together, but even so the bonding is very much weaker than the other bonds in the lattice, and therefore micas can be cleaved more easily along the unit planes than across them.

1.3.6 The Chlorites

Although often classified as clay minerals, the chlorites are structurally related to the micas. They consist of a substituted talc layer, of formula $Mg_3(Si_3Al)^-O_{10}(OH)_2$, similar to that in micas; this layer clearly carries a negative charge which is neutralized, not by a K^+ ion, but by a substituted brucite (magnesium hydroxide) layer, having the formula $Mg_2Al^+(OH)_6$. Because of the replacement of Mg by Al, this layer is positively charged, as shown. The unit formula of the chlorites is obtained by combining these two layers, to give the formula:

$$Mg_5Al_2Si_3O_{10}(OH)_8.$$

The chlorite crystal is made up of alternate layers of talc and brucite, as above. Naturally-occurring chlorites may differ somewhat from the above formula, because further substitutions may occur: e.g. a further Al may replace Si, or Fe may replace Mg or Al.

1.3.7 The Illites

Structure

There exists a group of minerals that is structurally similar to the micas, but contain less potash and more combined water than the latter. These materials are called *illites*, but have also been called hydrous micas or sericites.

The structure of the illites has not been completely elucidated, and it is not even certain whether they are a distinct species. It has been suggested that they have been formed from micas by the replacement of the K, Na or Ca by hydrated hydrogen ions (OH_3^+).

On the other hand, there is some evidence that they may be a closely-associated mixture of mica with a clay mineral.

Illites often occur as an accessory mineral in clays, to the extent of 40% or more.

1.3.8 Physical Properties of the Clay Minerals

Particle Size and Shape

Nacrite, *dickite* and *kaolinite* exist only as very small, hexagonally-shaped platy crystals, varying from less than 0.1μ to 2μ in diameter, although occasionally larger crystals of up to 20μ in diameter are found. Because they are so small they are best observed under the electron microscope, as are the tube-shaped particles of halloysite; these ultimate particles, as they are called, are quite hard and are not readily broken down by mechanical means, but they must not be confused with grains; a grain is a loose agglomerate of particles, caked together, that break up when they are mixed with water and agitated. Thus although grains of dry-ground fireclay may only just pass a 240-mesh standard sieve (72μ aperture), the ultimate particles are much smaller as can be shown by dispersing the clay in water and measuring the distribution of its particle sizes (see later).

The montmorillonites consist of thin, platy crystals, so small that their shape is difficult to discern, even with the electron microscope. Estimates of their size range from 0.01 to 2μ in diameter.

There is no doubt that the extreme smallness of clay particles is responsible for many of their intrinsic properties, such as plasticity.

Illites are very fine-grained minerals, particles as small as 0.05μ* in radius having been reported. They are said to be slightly plastic, although it is not certain whether this plasticity is due to contamination with clay mineral. Because of their alkali content, clays containing illite have a comparatively low refractoriness.

*1 μ (micron)= 10^{-6}m.

Specific Gravity

The specific gravity of the clays is about 2.6. It is difficult to measure the specific gravity of montmorillonite because it swells, but it is calculated that the value is close to that of the kaolins.

Effect of Heat on the Clay Minerals

Most hydrated minerals lose water when they are heated, and this, briefly, is what occurs when kaolinite, nacrite or dickite are heated above about 450°C:

$$\text{e.g.}\quad \underset{\text{Kaolinite}}{Al_2Si_2O_5(OH)_4} \xrightarrow{450^\circ} \underset{\text{“Meta-kaolin”}}{Al_2Si_2O_7+2H_2O}$$

When the water has been driven off, the residue still retains some of the crystalline features of the original kaolin mineral, and is therefore called “meta-kaolin”. Chemically, however, it behaves as if it were simply a mixture of finely divided silica and alumina.

Heating clay minerals to temperatures higher than 450° results in complex changes that are outside the scope of this work, but over about 1000°C the products are mullite and free silica, which may be represented by the equation:

$$\underset{\text{Meta-kaolin}}{3Al_2Si_2O_7} \xrightarrow{1000^\circ C} \underset{\text{Mullite}}{Al_6Si_2O_{13}} + \underset{\text{Silica}}{4SiO_2}$$

It must be borne in mind that many clays (as opposed to clay minerals) are impure, and these impurities can have a marked effect on the products formed and the temperatures of the various reactions. Nevertheless, the breakdown of kaolin minerals about 450°C is little affected by impurities, and it is this breakdown, which occurs during firing, that gives clay its unique properties. Before it breaks down, clay is sensitive to the action of water and can be moulded in the wet state; when the structure has been broken down by firing, and when the constituents have recombined at higher temperatures, the clay retains its shape well and is unaffected by water.

When montmorillonite is heated, water that has been absorbed between the silica layers is evolved first, between 100° and 200°C. About 700°C, the clay mineral breaks down, giving up its water of constitution, and an amorphous mass of silica, alumina and magnesia remains, corresponding to “meta-kaolin”. On being heated to about 1200°C, mullite, cristobalite, cordierite and spinel form.

All clays, on being heated to a high temperature begin to fuse, a viscous liquid being formed, which consists principally of the excess silica, together with various impurities such as Na_2O, K_2O, CaO and MgO. These oxides lower the melting point of the silica, enabling liquid to form (fusion) at the comparatively low temperature of 1200°C, and for this reason are called *fluxing oxides*.

On cooling, the liquid that was produced does not crystallize completely, but mostly solidifies to form a *glass*.

During firing, the liquid that forms when a clay begins to fuse fills up pore spaces in the clay; consequently the overall volume decreases, i.e. there is a *shrinkage*. Clearly the porosity decreases also and this reduction of porosity through the formation of liquid during firing is known as *vitrification*.

1.3.9 Clay–Water Suspensions

For casting with a fluid slip, a considerable proportion of water must be added to the clay which itself contains some moisture; for plastic mixes that have to be moulded, less water is added; finally for semi-dry pressing the moisture already in the clay may be sufficient.

1.3.10 Colloidal Properties

When a solid, powdered material is shaken up in a liquid and left to stand, the solid particles, provided that they are denser than the liquid, sink to the bottom of the container, forming a sediment. Under constant conditions, the rate at which particles settle (or sediment) depends on their size; the larger they are, the more rapidly they sink. For spherical particles of radius r m., density d_s, in a liquid of density d_l and viscosity η, the rate of fall v in m.sec^{-1} is given by Stokes's Law, which may be written:

$$v = \frac{2r^2(d_s - d_l)g}{9\eta}$$

g being the acceleration due to gravity.

For particles of any other shape, the value of r derived from Stokes' Law has no exact geometrical interpretation but is known as "the equivalent spherical radius" and is used as a convenient measure of the size of the particle. Stokes's Law is only true for particles greater than about 1μ in diameter, however; for smaller particles their natural thermal motion (called the Brownian Movement) becomes increasingly important. This movement causes small particles to diffuse upwards against the sedimenting force, and if they are sufficiently small they can prevent complete sedimentation. Such a suspension is said to be colloidal, and exhibits some very special properties that are not shown by coarse suspensions. Since a high proportion of clay-mineral particles are considerably smaller than 1 μ in diameter, they exhibit typical colloidal properties. Some idea of the relative sizes of particles in coarse suspensions, colloidal suspensions and solutions may be obtained from Table 3.

Table 3. Relative Size of Colloidal Particles

Type of particle	*Diameter* (μ)
Atoms and molecules	Less than 0.0005
Colloidal particles	0.001–1
Emulsions and suspensions	Greater than 1

Clearly, colloidal particles lie on the border-line between coarse particles, which obey the ordinary sedimentation laws, and atoms and molecules, which do not; it is therefore not surprising that colloidal particles behave as they do. Although we shall be concerned mainly with suspensions of solid particles in water, it should be realized that colloidal suspensions of solid particles in liquids other than water can be produced.

Properties of Colloidal Suspensions

Colloidal particles are invisible to the naked eye, and just beyond the limit of resolution of the optical microscope, but they can be detected indirectly by the light they scatter. If a beam of light is passed through a colloidal suspension, the light is scattered in all directions by the particles, and if a microscope is used to view the particles at right angles to the incident beam, the particles appear as bright points of light, although their shape cannot be made out. The phenomenon is known as the *Tyndall Effect*, and the arrangement for viewing the particles is called the *ultramicroscope* (Figure 16).

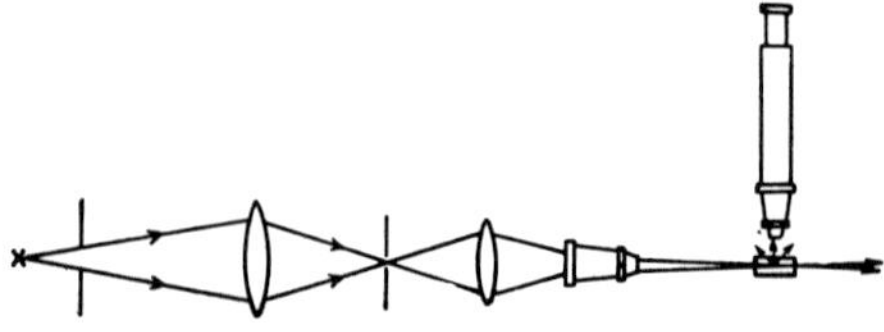

Fig. 16.—The ultramicroscope. (from Glasstone's Textbook of Physical Chemistry, copyright 1946, D. Van Nostrand Co. Inc., Princeton, N.J.).

It is by means of the ultramicroscope that the Brownian Movement may be observed.

Electrical Behaviour of Colloidal Suspensions

If two electrodes (preferably of platinum) are placed in a colloidal suspension, and a potential difference is applied (e.g. 200 volts from a battery or D.C. mains), the colloidal particles migrate to one of the electrodes, which shows that they are electrically charged (Figure 17). This phenomenon is known as *electrophoresis.* In a colloidal suspension of clay in water, the

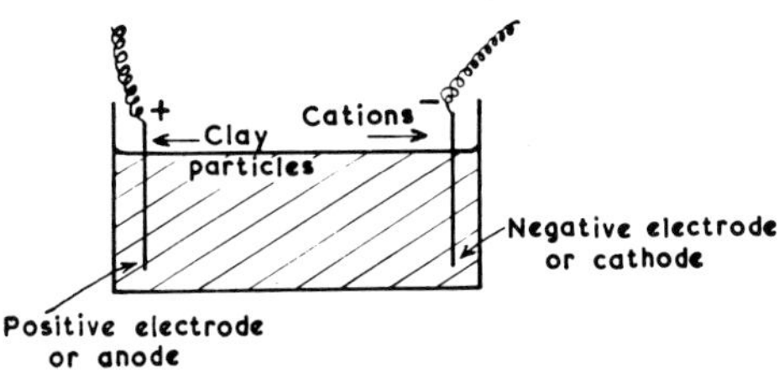

Fig. 17.—Electrophoresis.

particles are negatively charged and migrate to the anode. Now clearly the suspension as a whole must be electrically neutral, and therefore somewhere there must be an equal and opposite charge to balance it. In clay suspensions, this balancing charge is derived from positively-charged ions (cations) which migrate to the cathode (see Figure 17), give up their charges, and react with water to form hydroxides. This behaviour can be readily shown by testing chemically that part of the suspension close to the cathode; usually NaOH, KOH, $Ca(OH)_2$ and $Mg(OH)_2$ are present, showing that the corresponding ions, Na^+, K^+, Ca^{2+} and Mg^{2+} formed the "balancing charge" in the colloidal suspension.

Distribution of Charges in a Colloidal Suspension

So long as the balancing cations are not disturbed by electrophoresis, they remain close to the negatively-charged colloidal particles, forming a layer or "swarm" around them (Figure 18).

On the currently-accepted theory, some of the cations are held close to the surface of the clay particle as a strongly-adsorbed monomolecular layer (known as the Stern Layer); the remainder form a more diffuse layer in which the concentration of ions falls off exponentially with distance from the surface.

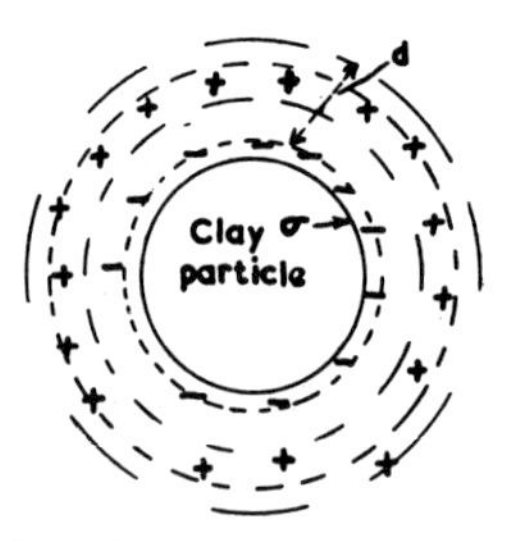

Fig. 18.—Distribution of charges on a colloidal particle of clay.

Approximately, the system may be considered as a spherical condenser, with two concentric charged plates, and it can be shown from elementary electrostatics that, if σ is the amount of electrical charge per unit area on the particle, *d* the effective distance apart of the positive and negative layers, and *D* the dielectric constant of the surrounding liquid (water), the electrical potential at the inner surface, ζ, is:

$$\zeta = \frac{4\pi d \sigma}{D} \qquad \text{(in S.I. Units, } \zeta = \frac{\sigma d}{kk_0}\text{)}$$

Although there is little doubt that the colloidal particle functions as a condenser, it will be clear that the distance *d* has no precise geometrical meaning but is regarded as an "equivalent distance".

In addition to ions, molecules of water, which act as electrical dipoles, are also attracted to the charged particle, and form a layer or sheath round it; this layer of firmly-held water molecules is known as the *lyosphere**. The lyosphere is limited in thickness because the electrostatic forces holding it fall off with distance from the particle. The particle, with its lyosphere, can be regarded as a single entity, called the *colloidal micelle.*

The use of the Greek letter ζ (zeta) to denote the effective potential of the system is now universally accepted, and it is usually spoken of as the *zeta potential.* The zeta potential is very important in determining the properties of colloidal suspensions, as we shall now see.

*Lyosphere is a general term applicable to all liquids; where water only is involved, the term hydrosphere is sometimes used.

1.3.11 Cation Exchange

It was pointed out earlier that the "balancing cations" associated with the negatively-charged particles consist of Ca^{2+}, Mg^{2+}, Na^{+} and K^{+} ions, but these ions are readily replaced by or exchanged for others, the reaction being very rapid. For example, if a natural clay is placed in a strong solution of ammonium chloride, NH_4Cl, ammonium ions, NH_4^{+}, tend to replace the existing ions in the outer layer, and the original balancing ions are released into solution as chlorides. The reaction can be represented by the equation:

$$\left.\begin{matrix}Ca^{2+}\\ Mg^{2+}\\ Na^{+}\\ K^{+}\end{matrix}\ \text{clay}\right| + NH_4^{+}Cl^{-} \rightleftharpoons NH_4^{+}\left|\text{clay}\right| + CaCl_2 + MgCl_2 + NaCl + KCl$$

Reactions of this type are known as cation exchange, sometimes incorrectly referred to as "base exchange". Note that the reaction is reversible.

It may be asked how the clay minerals possess this negative charge that requires to be balanced by exchangeable cations. One very obvious answer is isomorphous substitution. It will be remembered that, in the montmorillonites, charge deficiencies caused by such substitution have to be balanced by other cations which are outside the lattice and are therefore available for "exchange". For example, in the sodium form of montmorillonite, $Na_{0.33}(Mg_{0.33}Al_{1.67})Si_4O_{10}(OH)_2$, the sodium ion can be replaced by other cations. It is now believed that a small degree of substitution occurs in the kaolin group, which is responsible for the greater part of the cation exchange. The other possible cause of negative charges, particularly in the kaolins, is the presence of unsatisfied valencies at the edges of the crystals. It will be remembered that these sheet structures are capable of indefinite extension horizontally; since they must end somewhere, at the boundaries there will be O atoms that are unsatisfied, and appear as negative charges which require balancing by cations. This is the so-called "broken bond" theory.

For most types of cation, at least, the amount of cation exchange is a definite and constant quantity, called the cation exchange capacity, and is measured in milli-equivalents (m.e.) of exchangeable ion per 100 g of clay. For pure kaolinite, the cation exchange capacity varies from 2–12 m.e. per 100 g, whereas for montmorillonite it is about 100 m.e. per 100 g. This high value for the montmorillonites is due to the substitutions previously referred to. Disordered kaolinites (in ball clays and fireclays) have a high exchange capacity, up to 30 m.e., which may be attributable to substitution (see ball clays and fireclays).

1.3.12 Stability of Colloidal Suspensions

So long as a colloidal suspension remains as small discrete particles, it sediments extremely slowly, if at all, and is said to be in a *dispersed* or *deflocculated* state. If, however, the individual particles can be made to gather together to form aggregates they may become large enough for the normal sedimenting force to predominate, i.e. if large enough they will sediment according to Stokes's Law. The gathering together of particles into aggregates or "flocs" is called *flocculation.*

Now the zeta potential, recently described, is a measure of the energy of repulsion of the particles. As long as this potential is high (of the order of 50 mV or more) the particles repel one another so strongly that they do not come close enough to form flocs. Suppose, now, that the zeta potential is lowered (to say 10 mv or less); the particles can now come close enough to one another for the short-range forces of attraction to operate, and large flocs will be formed, causing the suspension to become unstable and to sediment.

The zeta potential, then, is the dominant factor in stability, and it is determined almost entirely by the type of exchangeable ions that are present on the clay. Referring to the equation for the zeta potential:

$$\zeta = \frac{4\,\pi\, d\sigma}{D}$$

it will be clear that if d, the interlayer distance, is large, then ζ will also be large; conversely, if d is small, ζ will be small. Some cations, notably H^+, Ca^{2+}, Mg^{2+}, and polyvalent cations generally, are mostly strongly adsorbed in the Stern Layer, so that the effective value of d is small and hence the ζ-potential is low. With the monovalent alkali ions Li^+, Na^+ and K^+, however, a much greater proportion remain in the diffuse layer and hence d is effectively high and so is the -potential (fig. 19).

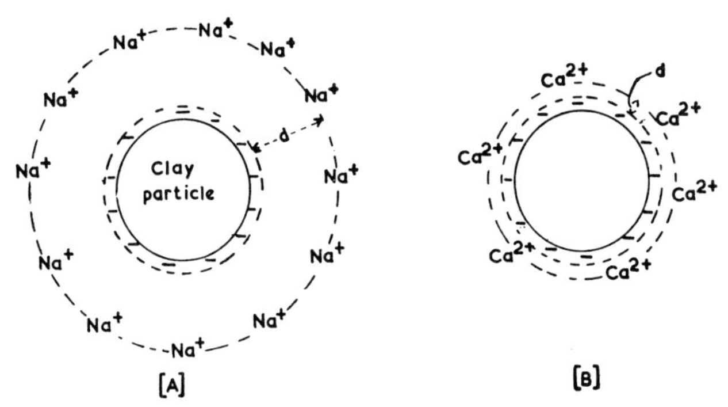

Fig. 19.—Distribution of charges in : A. Sodium clay. B. Calcium clay.

In practical terms, this behaviour means that the presence of alkali cations (Li^{+}, Na^{+} and K^{+}) causes suspensions to disperse or deflocculate whereas divalent cations (Ca^{2+}, Mg^{2+}, Ba^{2+}, Sr^{2+}) cause suspensions to flocculate. But if a large excess of any given cation is introduced, "crowding" of the double layer results, so that the two layers are "squeezed" together, reducing *d*, and therefore causing flocculation; this secondary cause of flocculation is sometimes called "salting-out".

1.3.13 Deflocculants

This "salting-out" must be borne in mind if we wish to apply our knowledge of ζ-potentials to the deflocculation of clay suspensions. Suppose, for instance, that we have a natural clay, the chief exchangeable ion of which is Ca^{2+}; initially a suspension of this clay is flocculated. If we wish to deflocculate it, we must clearly add a salt of sodium (or potassium), which contains these cations. However, if we add, say, sodium chloride, the reaction will be:

$$\text{Ca-clay} + \text{NaCl} \rightleftharpoons \text{Na-clay} + \text{CaCl}_2.$$

This is a reversible reaction, and so we do not obtain complete conversion to a Na-clay, because the $CaCl_2$ formed tends to reverse the reaction. In order to achieve complete conversion, we must get rid of the Ca^{2+} or at least convert it to a form in which it cannot interfere with the reaction. This is done by adding sodium carbonate:

$$\text{Ca-clay} + Na_2CO_3 \longrightarrow \text{Na-clay} + \downarrow CaCO_3.$$

The calcium is precipitated as insoluble calcium carbonate, which does not interfere, and so conversion is complete. The choice of deflocculant for maximum dispersion depends therefore on the type of exchangeable cations. Many deflocculants precipitate undesirable cations, as in the equation above, e.g. sodium carbonate, sodium oxalate. More complex ones, like sodium silicate, act in a slightly different way, as we shall see later.

1.3.14 Soluble Salts

If a natural clay is contaminated with soluble salts, they exert the "crowding effect", the clay flocculates and the addition of deflocculants alone has little effect. Moreover certain anions, notably SO_4^{2-}, are powerful flocculants. Fortunately in filter-pressing most soluble impurities are removed, but some remain and have to be removed chemically. Calcium sulphate, $CaSO_4$ (in the form of gypsum), is a common example of such a contaminant; if present, the best way of removing it is to treat the clay with the requisite amount of barium carbonate, according to the equation:

$$CaSO_4 + BaCO_3 = \downarrow BaSO_4 + \downarrow CaCO_3.$$

The barium carbonate is sufficiently soluble to initiate the reaction, and the final products, barium sulphate and calcium carbonate, are much less soluble than calcium sulphate and so do not interfere with deflocculation.

1.3.15 Adsorption

The surface of any solid is more active than its interior, and therefore tends to attract not only ions but whole molecules; this phenomenon is known as adsorption.

Owing to their fineness, clays possess a very great surface area per unit mass; e.g. a fine fraction of a fireclay has been found to have a surface area of 45 m^2 per g. Clearly clays are highly adsorptive; they all adsorb water on their surfaces, and this adsorbed water can only be completely removed by heating the clay to 105°C or higher, which shows the strength of the forces holding the water on the surface. Adsorbed molecules, whether of water or any other substance, tend to arrange themselves in a definite pattern, depending on how the electric charges within the molecule are distributed. Many such layers of molecules may be formed; it is believed that several molecular layers of water can be firmly attached to a particle of clay.

In general, the larger and more complex the molecule or ion, the more strongly it is adsorbed. Small anions, such as Cl^-, NO_3^- and SO_4^{2-}, are scarcely adsorbed at all, but large anions such as silicate (SiO_3^{2-}), tannate and polyphosphate are believed to be strongly adsorbed. When sodium silicate is added to a clay suspension, not only are the exchangeable cations replaced by Na^+, but silicate ions are adsorbed also, and this adsorption greatly increases the effective charge on the particle and so raises the zeta-potential still higher. Because of this two-fold action, sodium silicate is a very powerful deflocculant and only small proportions (of the order of 0·1–0·5% by weight of clay) are necessary for optimum deflocculation.

Clays take up dyestuffs, fatty acids and other organic substances partly through adsorption and partly through ion exchange. There are numerous practical applications of these reactions, but for further information the reader should refer to the reading list at the end of this book.

1.3.16 Applications of Colloidal Theory

The importance of clay-water mixtures in the ceramic industry has already been referred to. The next step is to determine how our knowledge of colloidal suspensions can be applied to industrial processes.

Measurement of Concentration

One of the most fundamental measurements that one has to make is the concentration of solid material (whether clay, flint, stone, or feldspar) in a suspension or slip. There are many ways in which this concentration can be expressed, but the method universally adopted in industry is that of slip density, expressed as the weight of slip per unit volume. In metric units this quantity will be in grams per ml., or in British units ounces per pint.* The conversion factor of 20 stems from the fact that 1 pint of water

*x g per ml. = $20x$ oz per pint.

weighs very nearly 20 oz and 1 ml. of water weighs very nearly 1 g at ordinary temperatures.

The potter frequently needs to know how much dry material is required to make a given volume of slip. Let us suppose that we have added m g of clay to w ml. of water, to make 1000 ml. of slip. Since the density of water=1, very nearly, the mass of the water is w grams, and therefore

the total mass is $(m+w)$ grams .... (1)

Now if the density of the clay is d g/ml., its volume will be m/d ml. The total volume of the slip is simply the volume of the clay plus the volume of the water, $(m/d)+w$, and this equals 1000 ml.

$$\text{i.e. } \frac{m}{d}+w=1000 \text{ ml.} \quad \ldots\ldots\ldots\ldots \quad (2).$$

If the density of the slip as a whole is S g/ml.,

$$\text{then } S = \frac{\text{Mass of slip}}{\text{Volume of slip}} = \frac{m+w}{1000}$$

$$\text{i.e. } m+w= 1000\,S \quad \ldots\ldots\ldots\ldots \quad (3).$$

But we see from Equation (2) that $w=1000—(m/d)$ and, substituting this value in Equation (3):

$$m+1000—(m/d)=1000S$$

Collecting the m's:– $m\left[1—(1/d)\right] = 1000S—1000 = 1000(S—1)$

whence $m=1000(S—1)/1—(1/d)=1000(S—1)\dfrac{d}{(d—1)}$.

Since the density d of most ceramic materials is roughly 2·5, the factor $d/(d—1)$ simplifies to 5/3, and so the weight of dry material per litre of slip=m=$1000(S—1)\times 5/3$. The weight of dry material for n litres would, of course, be $1000(S—1)\times 5/3\times n$. In British Units, m would be in ounces per pint of slip, S in ounces per pint, and the volume of slip would also be 1 pint. The equation then becomes: $m=(S—20)\times 5/3$ ounces per pint of slip, and in this form is known as Brogniart's formula.

The value for a slip from which pottery is to be cast, the slip density, or "pint weight" as it is often called, will vary from 32 to 42 oz. per pint.

Suspensions in general, and particularly those of clay, do not behave like true liquids, and the laws governing their flow are quite different. The laws of viscous flow are dealt with elsewhere and we shall only introduce some elementary ideas here, sufficient to understand the technology of casting slips and other clay-water systems.

The resistance that a liquid offers to flow, caused by internal friction between adjacent moving layers, is known as its *viscosity*. For a pure liquid, Newton defined the coefficient of viscosity as follows. Consider two parallel, adjacent planes in a liquid, of area A, distant x m. apart, moving relative to one another with a velocity v $m.sec.^{-1}$. The frictional force per unit area, F/A, is then given by:—

$$F/A = \eta \, dv/dx,$$

where η is the coefficient of viscosity. F/A is usually called the stress and dv/dx the velocity gradient or rate of shear. For pure liquids at constant temperature, η is also a constant. Suspensions, however, are very different; their viscosities are not constant, but are radically affected by mechanical disturbances, e.g. normal flow, shaking, stirring, etc. Strictly speaking therefore, the measured viscosity of a clay suspension is only an apparent one and should be called the "apparent viscosity". The viscosity of a clay slip depends, of course, on its concentration or density; the higher the density, the higher the viscosity. In industry it is usual to speak of the fluidity ($1/\eta$) rather than the viscosity of a suspension.

Whenever the apparent viscosity of a clay suspension has been altered by mechanical disturbance, it tends to return to its initial value when the disturbance is removed, and so many clay suspensions appear very viscous or even solid after standing for a time. If stirred vigorously, they appear to break down and flow very readily for a considerable time, but when left to stand, they gradually "thicken up" or become more viscous again, until eventually they regain their initial rigidity. This ability of certain clay suspensions to "thicken up" on standing, which is called *thixotropy*, is a very important property, and is particularly marked with slips that have been only partially deflocculated or that have been over-deflocculated. Slips that are deflocculated, especially with powerful deflocculants such as sodium silicate, possess very little thixotropy (Figure 20A). The thixotropy of a slip is, of course, also influenced by its density—other things being equal, the higher the density, the higher the thixotropy.

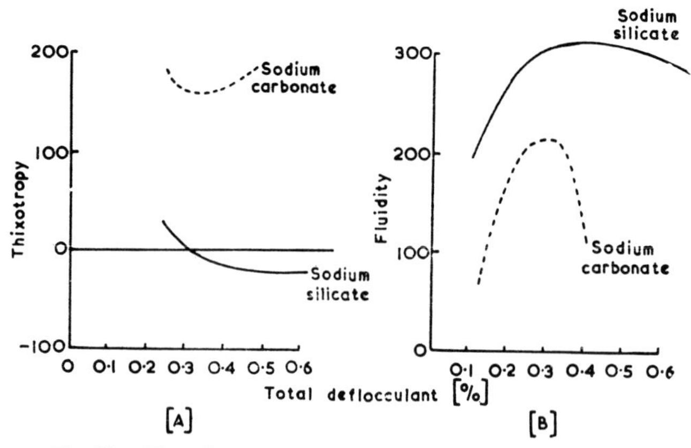

Fig. 20.—The effect of deflocculants on fluidity and thixotropy.

affected by deflocculants (Figure 20B). An untreated slip has a comparatively low fluidity, but when a normal deflocculant is added the fluidity increases up to a maximum (called the optimum value); with a further addition of deflocculant the value falls again (the region of over-deflocculation). Not all slips necessarily respond in the way shown in Figure 20, but the marked difference in the effects of sodium carbonate and sodium silicate are of fairly general applicability. The units in which fluidity and thixotropy are measured in Figure 20 are not "absolute" units, but are arbitrary units obtained from a torsion viscometer, which is described on page 40.

Flow Properties of Casting-slips

Figure 20B shows that an untreated slip would have a very low fluidity at the pint weight employed (35 oz/pint); and in fact without deflocculant it would be impossible to make a workable casting-slip at this pint weight. We could, of course, increase the fluidity by decreasing the density of the slip, but this procedure would prolong unduly the casting-time and poor casts would result. All casting-slips are therefore deflocculated.

For satisfactory casting a slip must have certain definite flow properties. No universal rules can be stated, but it has been found, for example, that sanitary earthenware casting-slips operate best with a fairly high fluidity and a moderate thixotropy. Since some thixotropy is required sodium silicate is rarely used alone as a deflocculant, but is usually added as mixture with sodium carbonate.

The casting-faults caused by incorrect fluidity and thixotropy depend to some extent on the type of the casting-slip, but some typical examples of such faults and their probable causes are quoted below (Table 4).

Table 4. Casting-faults for Sanitary Earthenware

Flow characteristic	*Casting-fault*
Fluidity too high	(1) Slow casting-rate (2) Cracking (3) "Wreathing"
Fluidity too low	(1) Pinholing (2) Bad draining
Thixotropy too high	(1) "Flabby" casts (2) Bad draining (3) Slow drying
Thixotropy too low	(1) Brittle casts (2) Difficulty in fettling (3) Slow casting-rate (4) "Wreathing" (5) Cracking

To correct such faults, let us suppose that the deflocculant is a 1:1 mixture of sodium carbonate and sodium silicate, and that the percentage of it to be added is always less than the "optimum"—below that required for maximum fluidity (see Figure 20). Under these conditions, the effect of adding water, clay or deflocculant is indicated in Table 5.

Table 5. Adjustment of Flow Properties of Casting-slips

Addition	*Fluidity*	*Thixotropy*
Water	Increase	Small decrease
Deflocculant	Increase	Decrease
Clay	Decrease	Increase

To alter any one flow property alone, leaving the others constant, it is clearly necessary to combine two of the three additions suggested in Table 5. For example, to increase the thixotropy of a casting-slip without altering the fluidity appreciably, it may be necessary to add untreated clay or body, and water. The addition of clay increases the thixotropy but at the expense of fluidity; the addition of water restores the fluidity to its original value, without unduly affecting the thixotropy. The most suitable values of fluidity and thixotropy vary from one slip to another and can only be found by trial and error.

Measurement of Flow Properties

For the precise scientific measurement of viscosity and thixotropy in absolute units, see "Rheology of Ceramic Systems". Here we shall restrict ourselves to a consideration of the torsion viscometer,

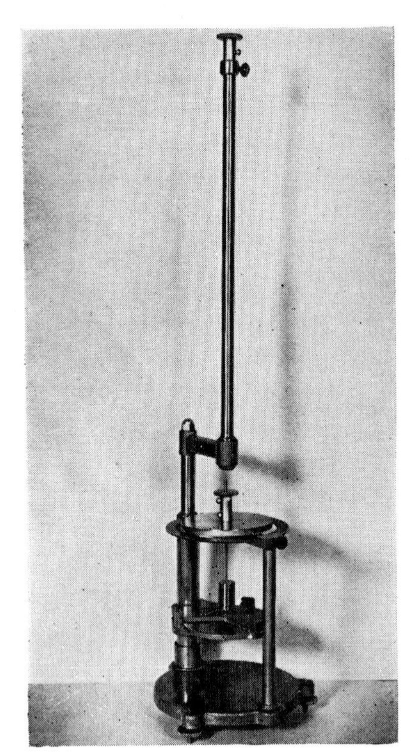

Fig. 21.—The torsion viscometer.

which is the instrument used most widely in the pottery industry (Figure 21). It consists essentially of a flywheel of standard size and mass, suspended by a phosphor-bronze torsion wire (of standard gauge) clamped from a torsion head. The flywheel carries underneath it a cylinder of given dimensions, and the rotation of the flywheel and cylinder is measured on a circular scale marked in degrees (0° to 360°). The base of the instrument has a movable platform, on which the sample under test is placed, and a set of screws for levelling.

Before a measurement is taken, the viscometer is levelled so that the cylinder, pin and flywheel hang freely. The torsion head is then adjusted until the pointer reads "zero" when at rest, and the instrument is "wound up" by rotating the flywheel one revolution in a clockwise direction and clamping it in this position. If released at this stage, the flywheel rotates in an anti-clockwise direction, and when the torque in the wire has been eliminated, it still has enough momentum to carry it through part of a second revolution, which is called the "overswing". (When correctly adjusted, the "overswing" in air (without any sample) is very nearly 360°).

First a sample of casting-slip in a small beaker or metal cup is stirred vigorously to break down any thixotropy. It is then placed, as quickly as possible, underneath the flywheel so that the metal cylinder is completely immersed in the slip. The flywheel (previously having been "wound up") is then released, and the amount of "overswing" is recorded. Clearly the lower the fluidity of the slip, the smaller this reading will be, because the viscous slip retards the cylinder. This reading measures the "fluidity" of the slip. Immediately after this reading has been taken, the viscometer is "rewound", as before and, exactly one minute after the first reading, the flywheel is released and a second reading is taken in the same way. This second reading will generally be lower than the first, because the one-minute period of rest enables some thixotropy to build up. The *fall* in the reading, therefore, is taken as the thixotropy; should there be an increase in the reading, the thixotropy is, of course, negative.

Dilatancy

Dilatancy is a property that is more common in coarse non-colloidal suspensions than in clays. A suspension is said to be dilatant when, being mechanically disturbed, it appears to stiffen but becomes mobile again when the disturbing force is removed. Suspensions of alumina and of quartz or sand are often dilatant, and some clays and bodies that contain a high proportion of non-plastic additives (flint, feldspar, etc.) may be also.

Dilatancy cannot be measured with the usual type of torsion viscometer found on a works but the more precise rotating-cylinder instruments can measure it. When very pronounced, it can be detected qualitatively by "feel".

Rheopexy

Rheopexy is another phenomenon exhibited by clays that contain much coarse material. Some clay suspensions that do not "stiffen" when left undisturbed can be made to do so by very gentle agitation or vibration; such clays are said to be rheopectic, since they exhibit rheopexy, which is sometimes called "induced thixotropy".

Permeability to Water

The pressing time for any clay under given conditions depends on its permeability to water, which in turn depends on the exchangeable ions. If they are of the flocculating type (e.g. H^+, Ca^{2+}, or Mg^{2+}) the clay is permeable to water and rapidly forms a firm press-cake. On the other hand, if the exchangeable ions are predominantly Na^+ or K^+, filter-pressing is difficult and the press-cakes are often "watery".

Fortunately, most natural clays are of the Ca^{2+} or H^+ type, and so press easily, but in some instances clay is a by-product (e.g. in the purification of fine coal by froth flotation) and during flotation is inevitably deflocculated, and before such clay can be pressed a flocculant (e.g. lime, calcium chloride) has to be added.

These phenomena and many others can be readily understood if we consider the state of dispersion of the clay. Consider (Figure 22A) a clay-water mixture in which the clay is flocculated (i.e. a H^+, Ca^{2+} or Mg^{2+} clay.) The ultimate particles are not independent but are bunched higgledy-piggledy. As water is removed, whether by drying, filter-pressing or other means, the particles come closer together, and eventually they pack closely, as in Figure 22A. However, owing to the initial disorder of the flocs, their packing is very loose, with a high proportion of voids; such a body is clearly very permeable by water, and so it can be filter-pressed satisfactorily.

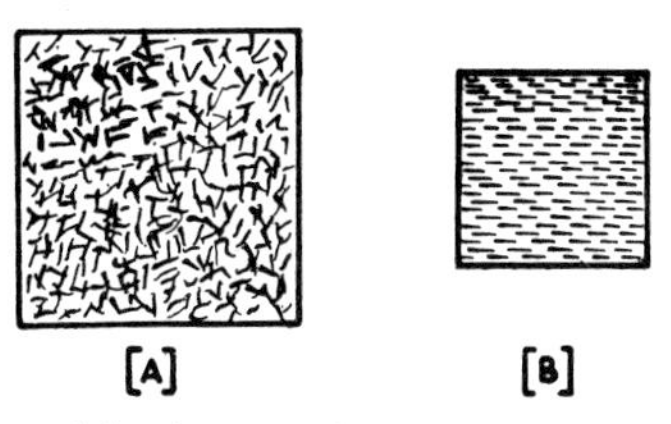

Fig. 22.—Packing of particles in : A. Flocculated clay. B. Deflocculated clay.

Consider now a deflocculated clay (Figure 22B). Here the ultimate particles are prevented by the high zeta-potential from agglomerating and so remain as separate platelets. When water is removed from such a system, the particles draw closer together, and eventually, being plate-like, they pack one upon the other in an ordered fashion, as in the illustration. In this system the packing is clearly very dense, consequently the system is not easily penetrated by water, with the result that filter-pressing is difficult.

Dry Strength

For the reasons given above, flocculated clays are, on the whole, easier to dry than deflocculated clays.

The strength of dry unfired clays, however, depends principally on the area of contact between adjacent particles. From Figure 22 it is evident that the deflocculated system B has a greater area of contact than has system A, which explains why deflocculated clays have a higher dry strength than flocculated clays.

1.3.18 Plasticity

Although plasticity is a recognizable property, it is difficult to define scientifically. A substance is said to be plastic if it is cohesive (i.e. holds together), is mouldable to any shape, and retains that shape. If we restrict ourselves to solid–water mixtures, we find that there are few substances other than the clay minerals that when mixed with water develop plasticity, and they all have at least one thing in common—their ultimate particles are very small—of colloidal dimensions.

We have already mentioned that colloidal particles are charged; being charged causes water molecules to be attracted to them, forming a coherent film, and probably accounts for the cohesiveness of clay–water mixtures. The surface tension of the water also assists cohesion, while at the same time the charged particles resist being squeezed too closely together by their mutual repulsion.

The factor contributing most to plasticity is undoubtedly the effective surface area of the ultimate particles of clay, which of course depends on their fineness, consequently the smaller the ultimate particles of a clay, the more plastic it is likely to be. The type of the exchangeable ions will also affect plasticity—but the factors that are involved conflict and so it is not possible to decide which ions are responsible for the highest plasticity. For the same moisture content, Na- and K-clays deform more easily than Ca- or H-clays, but with the former the moulding-pressures are more critical.

Measurement

Many attempts have been made to develop a method for measuring plasticity but so many factors are involved that no single method is entirely satisfactory. Only two methods will be described here; they have been chosen because they are easy to understand and simple to carry out.

The first method really provides a measure of the cohesiveness or bonding power of the clay. To a stiff clay-water mixture a non-plastic material, such as graded sand, is added a little at a time, until the clay–sand–water mixture falls apart. The percentage of sand (based on the dry weight of the clay) necessary to cause this is the "plasticity index" or bonding power of the clay.

In the second method, developed by Atterberg, a thick suspension of the clay is made up, and poured on to a plaster slab. As the plaster absorbs water from the clay, the clay becomes stiffer and eventually reaches a stage at which it will no longer flow under its own weight. To test from time to time a knife is thrust into the clay until the mark it makes no longer "heals"; the moisture content of a sample of the clay in this condition is then determined, and is known as the "upper plastic limit". When the clay has dried further, it is removed from the plaster slab and worked in the hands. This working dries the clay further by evaporation, and eventually it loses its cohesiveness and, when pressed between the fingers, crumbles. The moisture content is then again determined and this value is called the "lower plastic limit". The arithmetical difference between the upper and the lower plastic limit is the plasticity index of the clay, P:

$$P = \text{(Upper plastic limit)} - \text{(Lower plastic limit)}.$$

Both moisture contents are calculated on the basis of 100 parts of dry clay.

1.4 NATURAL CLAYS

In the foregoing pages we have been concerned with pure clay-minerals. We now turn to natural clays—clays as they are actually found in the earth. In addition to the clay mineral, a natural clay contains other minerals as impurities, which naturally affect the physical properties of the clay. It is therefore important to know what these impurities are and how much of them is present—the mineralogical composition of the clay.

An ultimate chemical analysis alone does not give all the necessary information. For instance, if an analyst reports that a clay has a silica content of 60%, we do not know whether this silica is present as "free silica", quartz, or combined as kaolinite or as mica. If, however, we can find out precisely what minerals are in the clay, it may be possible to calculate the percentages of each mineral from the chemical analysis and this calculation provides the so-called "rational analysis" of the clay.

In order to determine the minerals in the clay, one has to resort to X-ray analysis, thermal analysis, and microscope examination. A number of samples of one type of clay, e.g. ball clay, having been examined by these means, it is fair to assume that other ball clays contain the same minerals and to calculate the "rational analysis" on this assumption. In certain instances, the percentage of one or more minerals can be estimated direct from the X-ray and thermal analyses, but it is doubtful whether the results are more accurate than those of the "rational analysis".

If a clay contains many minerals, the calculation of the rational analysis can become very involved, or even impossible, according to the minerals concerned and the completeness of the chemical analysis. We shall restrict ourselves to the rational analysis of kaolin-type clays that contain the following principal impurities:

Quartz	SiO_2
Potash mica (muscovite)	$KAl_3Si_3O_{10}(OH)_2$
Soda mica (paragonite)	$NaAl_3Si_3O_{10}(OH)_2$

Fortunately this description covers most of the commonly-used British clays—ball clays, fireclays, china clays and some brick-clays. For clays containing a montmorillonite or a chlorite, rational analysis presents a much more difficult problem, requiring special analytical methods.

1.4.1 Method of Calculation

Let us consider a typical ball clay, containing kaolinite as the principal clay mineral, with quartz and mica as the major impurities (Table 6). It is simpler to begin with the alkalis, since

Table 6. Chemical Analysis of a Ball Clay

Oxide	*Weight* (%)
SiO_2	51.13
Al_2O_3	29.30
Fe_2O_3	2.78
TiO_2	2.08
CaO	0.15
MgO	1.68
Na_2O	0.12
K_2O	2.80
Ignition loss	9.54
Total	99.58

they are present solely as micas*. Taking two "unit formulae" of soda mica, which can be written as $Na_2O.3Al_2O_3.6SiO_2.2H_2O$, it will be seen that they contain one "molecule" of Na_2O. Adding up the atomic weights of the empirical formulae of soda mica and sodium oxide respectively, we get:

$$\underbrace{Na_2O.3Al_2O_3.6SiO_2.2H_2O}_{764.21 \text{ parts by weight}} = \underbrace{Na_2O}_{61.99 \text{ parts by weight}}$$

From this we deduce that 1 part by weight of Na_2O corresponds to $\frac{764.21}{61.99}$ or 12.33 parts by weight of soda mica. From Table 6 the percentage of Na_2O in this clay is 0.12%, and the percentage

*Strictly speaking, a correction should be made for any alkali present as exchangeable ions.

soda mica is therefore $0.12 \times 12.33 = 1.48\%$ (to 2 decimal places).

In exactly the same way, from the unit formula of potash mica we find that 1 part by weight of K_2O corresponds to 8.46 parts by weight of potash mica. Hence from Table 6 the percentage of potash mica in this clay is $2.80 \times 8.47 = 23.72\%$.

In the next step of the calculation, we must remember that the alumina (29.30% from Table 6) is combined in three minerals—kaolinite, soda mica, and potash mica. Knowing now the percentage soda and potash micas, we can calculate from their formulae how much Al_2O_3 they contain. Subtracting this amount from the total Al_2O_3 (29.30%), we can find how much Al_2O_3 is present as kaolinite, and hence calculate the percentage of kaolinite.

The calculated percentage of Al_2O_3 in soda mica is 40.02%—i.e. 1 part by weight of soda mica contains 0·4002 parts by weight of Al_2O_3, and so the percentage of Al_2O_3 combined as soda mica in our clay$=1.48 \times 0.4002 = 0.59$ (to 2 decimal places).

Similarly, the percentage of Al_2O_3 in potash mica is 38.40%, and the percentage of Al_2O_3 combined in potash mica in the clay$=$ $23.72 \times 0.3840 = 9.11\%$. Hence the total Al_2O_3 combined in the form of micas $= 0.59 + 9.11 = 9.70\%$.

Therefore the remainder of the Al_2O_3 (that combined as kaolinite) $= 29.30 - 9.70 = 19.60\%$.

But 1 part by weight of Al_2O_3 corresponds to 2.532 parts by weight of kaolinite ($Al_2O_3.2SiO_2.2H_2O$). Therefore the percentage kaolinite in the clay $= 19.60 \times 2.532 = 49.63\%$.

Finally, the quartz or "free silica" has to be calculated. The total SiO_2, which from Table 6 is 51.13%, consists not only of quartz, but of silica that is combined in the micas and in the kaolinite. Therefore first we must calculate the amount of "combined" silica and subtract it from the total, in order to arrive at the "free" silica or quartz.

To eliminate further lengthy calculations, the factors by which the micas and the kaolinite have to be multiplied in order to deduce the amount of SiO_2 in them are set out in Table 7, which also includes the other factors used in the previous part of the working. In using Table 7, we look under the first column for the substance that we wish to estimate. We then look in the second column for the substance from which it is to be calculated, and immediately opposite, in the third column, is the conversion factor. Thus to

Table 7. Conversion Factors for Rational Analysis

Sought :	*Found as :*	*Conversion factor*
Soda mica	Na_2O	12.33
Potash mica	K_2O	8.47
Alumina	Soda mica	0.400
	Potash mica	0.384
Kaolinite	Al_2O_3	2.532
Silica	Soda mica	0.471
	Potash mica	0.452
	Kaolinite	0.465

find proportion of silica combined as soda mica, we find "silica" in column 1, and opposite "soda mica" (column 2) we find the conversion factor (0.471) in column 3. Thus, silica combined as soda mica

=(Calculated amount of soda mica)×(Conversion factor)
=1.48×0.471
=0.72%.

Similarly, silica combined as potash mica
=23.72×0.452=10.72%.

Again, from Table 7, silica combined as kaolinite
=49.63×0.465=23.08%.

Total combined silica=0.72+10.72+23.08=34.52%.
From the analysis (Table 6), total silica=51.13%.
"free" silica or quartz=51.13—34.52=16.61%.

The final results are given in Table 8. Note that the principal minerals add up to only 91.4%; the remainder is assumed to be miscellaneous oxides, carbonates, or organic matter.

Table 8. Calculated Rational Analysis of a Ball Clay

Mineral	*Weight (%)*
Soda mica	1.5
Potash mica	23.7
Kaolinite	49.6
Quartz	16.6
Miscellaneous oxides (Fe_2O_3, TiO_2, CaO, MgO and organic matter)	8.6

1.4.2 Errors in Rational Analysis

Although the results in Table 8 are given to the first decimal place, it is doubtful whether the method warrants such accuracy, for the errors arise, not in the calculation, but in the assumptions that are made. For example, the composition of potash mica does not always correspond exactly to the "ideal" formula, $KAl_3Si_3O_{10}(OH)_2$, and therefore the conversion factor of Table 7 may differ slightly from the stated value. The same thing applies to other minerals. Furthermore, the clay may contain silicate minerals other than kaolinite, micas and quartz; this would render the method of calculation unsound. In cases of doubt, however, the composition should be checked by thermal analysis, X-ray analysis, or both.

A good deal of research needs to be done on the composition variations of pure minerals. As will be seen later, ball clays and fireclays contain a "disordered" kaolin mineral that departs considerably from the "ideal" formula, $Al_2Si_2O_5(OH)_4$. This obviously causes further inaccuracy in the calculation of rational analyses.

1.5 GEOLOGY OF THE CLAYS

1.5.1 Rocks

Any compacted earth is known to the geologist as a rock, and this broad definition includes substances such as coal and clay, despite their chemical dissimilarity. There are two main types of rock, *igneous* and *sedimentary*. Igneous rocks are formed by the solidification of molten material, called magma, from the hot interior of the earth. Owing to the high temperatures and continual disturbances in the earth's interior, molten material has been forced to the surface at various periods in the earth's history, and has solidified to form igneous rocks of various ages.

Sedimentary rocks are formed when igneous rocks are decomposed by various agencies, which can be classified as *epigenic* or *hypogenic*. *Epigenic agencies*, which include running water, carbon dioxide, winds, and glaciers, slowly decompose igneous rocks and break them into very small fragments, which are carried down to the river estuaries, lakes or seas, and deposited on the seabed. These deposits later form sedimentary rocks, of which clay is a typical example. The deposits do not consist only of clay, because some parts of an igneous rock decompose more slowly than others, and may form another kind of deposit, lying on top of the first. Other products of epigenic action are soluble, e.g. calcium and magnesium bicarbonates, colloidal silica, and sodium and potassium salts. As the water became saturated with some of these salts, they were precipitated, forming beds of limestone, magnesian limestone and siliceous rocks. As each layer became submerged under others, it was subjected to increasing pressure and so consolidated into a hard mass of rock.

It might be thought a simple matter to determine the relative age of sedimentary deposits, since the older deposits lie deeper than the more recent ones. But the deposition of sediments was periodically interrupted by earth movements, resulting in folding and upheaval of the deposited layers, and often either causing sea-beds to be raised above the water level or submerging land masses beneath the ocean. As a result of these disturbances, some deposits were washed away by newly-formed rivers, and some older deposits now appear immediately below the soil and are said to "outcrop".

Geologists have been able, nevertheless, to determine the "age" of these deposits, partly from the order in which they appear, partly from the fossils that are present and, more recently, by tests of radioactivity (Table 9). The alteration of igneous rocks by hot gases and water vapour from the interior of the earth is called *hypogenic action.*

1.5.2 Composition of Igneous Rocks

The kinds of mineral formed when molten material cooled to form igneous rocks depended on the initial composition, the temperature at which the material solidified, and the pressure. Generally speaking, the proportion of basic oxides increases with depth—in fact the core of the earth is thought to be very rich in iron. Consequently the igneous rock known as *granite* is considerably less basic than *basalt*, which originated from lower down in the earth's crust.

Granite and *basalt* are the commonest igneous rocks; there are many others, but their mineralogical compositions are similar. The chief consti'uents of granite, and basalt and their decomposition products are shown in Table 10.

Table 9. The Geological Systems

Era	*Period*	*Approximate age (Millions of years)*	*Principal clay deposits*	*Other deposits*
Lower Palaeozoic (or Primary)	Cambrian	500		Sandstones, slates, siliceous grits
	Ordovician	410		Sandstones, slates
	Silurian	350		Slaty and calcareous shales, limestones, flagstones, sandstones
Upper Palaeozoic (or Primary)	Devonian	325	Brick shales	Old Red Sandstone, and limestones
	Carboniferous	285	Fireclays, Etruria Marl	Mountain limestone, millstone grit, Lower, Middle and Upper Coal Measures
	Permian	210	Red and White marls	Magnesian limestone, gypsum and sandstones
Mesozoic (or Secondary)	Triassic	170	Keuper Marl	Bunter sandstone, pebble beds, Keuper Sandstone, gypsum
	Jurassic	145	Oxford clay, Fuller's earth	Limestone, gypsum, calcareous sands and grit, jet and lignite

Table 9. The Geological Systems—*cont.*

Era	*Period*	*Approximate age (Millions of years)*	*Principal clay deposits*	*Other depostis*
MESOZOIC (or SECONDARY)	CRETACEOUS	120	Gault clay, Hastings clay, Weald clay	Upper and Lower Greensands, Chalk, flints
CAINOZOIC (or TERTIARY)	EOCENE	60	Dorset ball clays, London clay, Reading beds, Bracklesham Beds.	Pebbles, sands, lignite
	OLIGOCENE	35	Devon ball-clays	Siliceous limestone, lignite
	MIOCENE	20		
	PLIOCENE	8		Calcareous sands, loams and limestone
QUATERNARY	PLEISTOCENE	1	Alluvium brick-earths	Gravel, chalk, flinty shingle, sands
	HOLOCENE (or RECENT)	—	Glacial clays, Boulder Clays	Sand and gravel, alluvial and glacial drift

Some igneous rocks, e.g. granite, contain altered material (formed by hypogenic action) that has remained embedded instead of having been removed to form sediments. Important examples are the china stones of Cornwall, which have been partially altered to form kaolinite and subsidiary products. The hypogenic agents responsible for this alteration were probably water vapour and gaseous compounds of fluorine and boron at high temperature, which may explain why fluorine- and boron-bearing minerals, such as tourmaline and fluorite, are found associated with the china stones.

Table 10. The Composition of Igneous Rocks and their Breakdown Products

Mineral	*Approximate formula*	*Approximate % present*		*Probable decomposition products*
		Granite	*Basalt*	
Orthoclase	$KAlSi_3O_8$	70 (Orthoclase and Anorthite)	10 (Orthoclase and Anorthite)	Kaolinite, colloidal silica, K_2CO_3
Anorthite	$CaAl_2Si_2O_8$			Kaolinite, colloidal silica, $CaCO_3$
Quartz	SiO_2	25	—	Unchanged
Hornblende	$(Na,K)_2(Fe^{2+},Mg)(Fe^{3+}, Al)_4\ Al_2Si_6O_{22}(OH)_2$	—	90 (Hornblende, Pyroxene and Olivine)	Kaolinite or montmorillonite, limonite, haematite, $CaCO_3$, $MgCO_3$, colloidal silica
Pyroxene	$(Mg,Fe)SiO_3$	—		Colloidal silica, limonite, haematite, $MgCO_3$
Olivine	$(Mg,Fe)_2SiO_4$	—		As for pyroxene
Muscovite	$KAl_3Si_3O_{10}(OH)_2$	5 (Muscovite and Biotite)	—	Probably unchanged
Biotite	$K(Mg,Fe)_3Si_3AlO_{10}(OH)_2$		—	Kaolinite or montmorillonite, iron oxides, colloidal silica, $MgCO_3$, K_2CO_3

1.5.3 Composition of Sedimentary Rocks

From Tables 9 and 10 it is evident that clay minerals are one of the important constituents of sedimentary rocks. With repeated deposition and submergence, the layers (or seams) of deposited clay were compacted under considerable pressure, and became quite hard and "rock-like". If the pressure was high enough, they may have formed *shales* or *slates*—laminated rocks of clay.

The type of clay mineral formed, however, depended on the climatic conditions prevailing at the time (this bears no relation to present-day conditions). Under fairly warm, humid conditions, kaolinite appears to have been formed, whereas under cool and somewhat drier conditions, montmorillonite seems to have been the end-product.

In addition to clays, many other minerals were formed. Sparingly-soluble salts such as calcium and magnesium carbonates and colloidal silica, were precipitated from sea-water and deposited on the sea bed, forming limestone, magnesian limestone, and silica rocks.

Throughout the ages, tropical trees and plants were abundant, particularly in the Carboniferous Period; as they decayed they became submerged, and through bacterial and chemical agencies were converted into either lignite (brown coal), bituminous coal or anthracite, depending on physical conditions. These organic deposits are now found associated with clay seams—lignite with the ball clays, bituminous coal and anthracite with the fireclays. In some way, part of the organic matter has become adsorbed on to the clay and cannot be separated from it; it is probably very finely divided, and can markedly affect the physical properties of clays.

1.5.4 Occurrence of Deposits

A geological map of Great Britain shows that the sedimentary deposits have been much disturbed by earth movements and are not always to be found as a regular sequence (Figure 23). In any given locality, some intermediate layers are found to have been washed away before further deposition occurred, so that they are now missing from the sequence. Continued folding of the earth's crust has tilted many sedimentary deposits out of the horizontal, so that many comparatively old deposits, like the Carboniferous, outcrop. This outcropping of deposits, particularly of coal, is

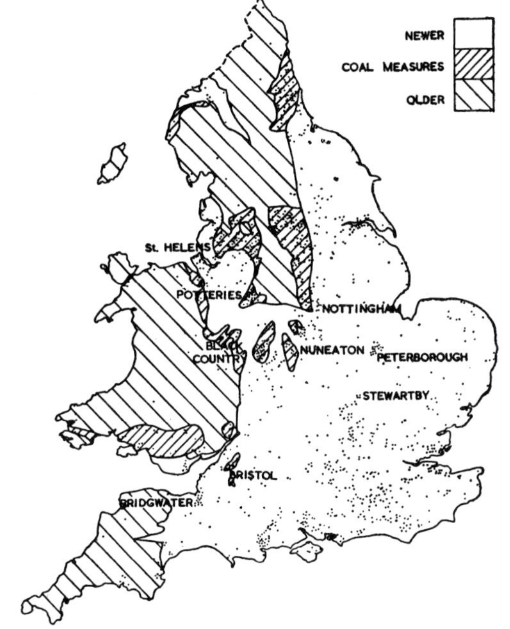

Fig. 23.—Simplified geological map of England and Wales.

important economically, otherwise many coal and clay deposits would be inaccessible. Even so, outcrops are usually covered with a thin layer of soil, called the "overburden" which in open-cast mining is often removed so as to expose underlying seams.

1.5.5 Classification of Clays

Any attempt to classify clays logically would result in a good deal of confusion, because such a classification would inevitably cut across many of the present well-recognized groups. In this section, therefore, we shall endeavour to make the best of the existing classification, explaining the meaning of each group-name as it is understood today.

The geologist recognizes two main types of clay, *residual* and *sedimentary*. *Residual* clays are those which have not been transported by natural agencies, but have remained in their place of origin, such as the Cornish china clays, that are found with the granite rock from which they were formed by hypogenic action. *Sedimentary clays* are those that have been removed from their original source by streams, glaciers, etc. Ball clays, fireclays and brick-clays, and in fact the majority of British clays, are of this type. Owing to mechanical abrasion during transportation, the particles of sedimentary clays tend to be smaller than those of residual clays. Moreover, sedimentary clays contain impurities that have been picked up during transportation.

1.5.6 Ball Clays

The ball clays are sedimentary deposits that were laid down in the Eocene and Oligocene Periods, and so called because they were originally dug out of the ground in blocks or "balls". They are an important raw material and are marketed both in the United Kingdom and abroad. Careful examination of a large number of samples of ball clay has established that their composition varies widely and so we cannot distinguish ball clays by their composition but only by their location and geological age.

The areas in which ball clay is worked are fairly small, and are confined to: the valley of the Bovey and Teign, near Newton Abbot; a depression near Torrington in North Devon; near Wareham and Corfe Castle in the Isle of Purbeck; and also near Wimbourne, Dorset. (See Figure 24).

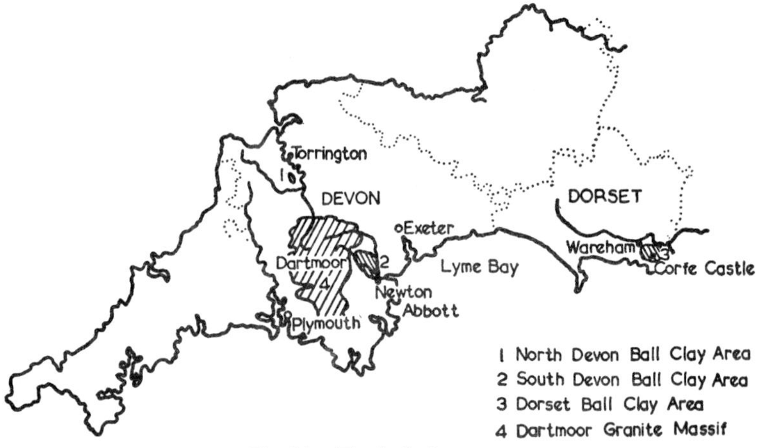

Fig. 24.—The ball clay areas.

Extraction

The method adopted for extracting sedimentary clays depends on the depth of the clay seam, the hardness of the clay, the depth of "overburden", and the inclination of the seam to the horizontal. *Deep mining* is usually only economic if the clay is very valuable and if some mineral, e.g. coal, is being extracted with it. *Open-pit methods* are employed when the seam is not too deep.

Ball clays are mined either by "open-pit" or underground methods, depending on the depth of the deposit.

For open-pit working, the overburden of soil is removed by mechanical excavators or bulldozers, and then the underlying clay is dug out by means of pneumatic shovels and loaded on to lorries by means of an elevator (Figure 25).

Fig. 25.—Operators cutting out ball clay with compressed-air-operated spades, (photograph by courtesy of English Clays, Lovering Pochin & Co. Ltd.).

Where underground mining is employed, the workings seldom extend to more than 120 ft below ground, the clay being hauled directly to the surface by mechanical excavator. No horizontal headings radiate from the central shaft, as they do in coal mining.

Chemical Composition

Table 11 illustrates the wide variations in chemical composition.

Table 11. The Variation of Chemical Composition of Ball Clays

Oxide	*Range of variation* (%)
SiO_2	40–60
Al_2O_3	25–40
Fe_2O_3	0.25–4.0
Na_2O	0–0.75
K_2O	0.5–4.0

If the so-called "siliceous" ball clays are included, the maximum SiO_2 content rises to 80%, and the minimum Al_2O_3 content falls to 15%. Outside these limits the deposits cease to be regarded as clays and probably fall into the category of sands or loams.

The principal clay-mineral of ball clays is of the kaolin type, and closely resembles kaolinite but is less well-crystallized, and its empirical formula departs slightly from the ideal, approximating to:

$$(Ca_{0.05}Fe^{3+}_{0.1}Mg_{0.1}Al_{1.8})Si_2O_5(OH)_4$$

The Al is slightly less than 2 atoms per unit formula, small proportions of other ions taking its place. In addition to the clay mineral, mica, quartz, miscellaneous oxides and carbonaceous matter are present.

Ball clays are variously described as "blue", "black" or "ivory", according to their colour: the blueness or blackness is caused by the presence of much organic matter whereas ivory clays owe their colour to the presence of iron oxide. The colour of an unfired ball clay, however, is not necessarily any guide to its fired colour.

Particle Size Distribution

Since the particles of ball clays cover a wide range of sizes, the best methods of representation are either:

(1) To specify the percentage of material falling within a given restricted range (if done graphically the result is known as a histogram) or

(2) To plot graphically the percentage of material less than a given size against that size. As the selected size is increased,

clearly the percentage of material less than this size must increase and must eventually reach 100%, when a limiting size is reached. For this reason, the resulting graph is known as a cumulative distribution curve.

Table 12 indicates the average size distributions of a number of ball clays, grouped according to their locations. Although 70–80% only of the total clay is accounted for in Table 12, the remainder is, of course, material larger than 1.0μ and consists chiefly of coarse-grained quartz, mica, coaly matter and other impurities.

Table 12. Average Particle-size Distribution for Ball Clays

Source	*Percentage material within the following size-ranges (radii in μ)**						
	<0.05	0.05–0.10	0.10–0.25	0.25–0.50	0.50–0.75	0.75–1.0	*Total accounted for*
North Devon	12.0	15.4	24.3	12.5	5.5	4.7	74.4
South Devon	9.6	15.7	29.2	17.9	5.8	3.7	81.9
Dorset	15.7	19.5	24.7	13.1	5.2	2.9	81.1

* = 1 micron (μ) = 10^{-6} m.

Evidently it is not sufficient simply to state that one clay is "finer" than another; it is necessary to specify the range of particle sizes for which this statement is true. For example (Table 12) the South Devon clays contain 81.9% material less than 1.0μ in radius—slightly more than the Dorset clays (81.1%), and on this basis the former might be considered the "finer" of the two. But the Dorset clays contain much more material in the ranges <0.05μ and 0·05–0·10μ; since these lower ranges are more important for colloidal behaviour, the Dorset clays may well be considered to be the "finer".

Specific Surface Area

A property closely related to particle size is specific surface area, which is considered to be the total area of the surfaces of all the particles of clay in unit weight of that clay, and is usually expressed in square metres per gram.

Clearly the smaller a particle, the greater its specific surface area; for example, if we take one solid piece of any material having a mass of 1 g, and cut it into smaller pieces, with each cut we expose fresh surfaces so that the total surface area increases although the total mass remains the same, and so the finer fractions of a clay contribute more (per unit weight) to the surface area than the coarser ones. The specific surface area of a clay can be calculated from the size distribution, but this method is unreliable, and better methods are based on the adsorption of mono-molecular films of gases (or solids from solution) on to the clay.

Wet-to-Dry Shrinkage

The wet-to-dry shrinkage of ball clays is high. In a survey of ball clays, values of linear shrinkage of up to 15% were quoted with leatherhard moisture contents* ranging from 10 to 32%, but in practice the effective shrinkage of ball clay in pottery bodies is considerably reduced by the other ingredients. A ball clay suitable for pottery manufacture should have a linear shrinkage of not more than 12%.

The greater the proportion of colloidal material in a clay, the greater the shrinkage. Moreover, Ca^{2+}, Mg^{2+} and H^+ clays shrink more than the corresponding Na^+ or K^+ clays.

Dry Strength

Ball clays are noted for their high dry strength, which is why they are used in pottery bodies and to a small extent in china. The dry strength of a ball clay depends on the proportion of "clay substance", on its fineness, on the exchangeable cations, and on the amount of organic matter present. Generally H^+, Ca^{2+}, and Mg^{2+} clays have lower dry strengths than Na^+ or K^+ (see page 43). As might be expected, the greater the proportion of "clay substance" and the greater the proportion of "fines", the higher the dry strength. Values of up to 1000 lb/in^2 have been recorded for the cross-breaking strength of very plastic ball clays, and those clays that contain much organic matter (e.g. the black ball clays) tend to be the strongest.

Base Exchange and Deflocculation

Ball clays have a high cation-exchange capacity, ranging from about 5 to 20 m.e./100 g, and so many of them require a con-

*Leatherhard moisture content is the moisture content at which a drying plastic clay body ceases to shrink.

siderable amount of electrolyte for deflocculation, particularly those with a high percentage of organic matter; the latter are, moreover, not readily overflocculated. The chief exchangeable ions are H^+, Ca^{2+}, Mg^{2+}, Na^+, and K^+.

The general principles of deflocculation have already been described but it should be remembered that the relative amounts of the exchangeable ions considerably affect the response to a given deflocculant. For instance, clays usually respond well to a 1:1 mixture of sodium carbonate and sodium silicate, but this ratio may have to be altered if the distribution of exchangeable ions is abnormal

Soluble Salts

If more than about 0.5% of soluble salts is present, adequate deflocculation is difficult to achieve; fortunately most ball clays contain less than 0.2% of such salts.

Fired Colour

It is important in the manufacture of pottery that the clays are "white-firing"—when fired they should be just "off-white" or pale cream, and most ball clays fulfil this requirement. The South Devon clays are reported to fire whitest ("off-white" to "creamy-white"); the North Devon clays fire "pale ivory" to "ivory", and those of Dorset fire "ivory", buff or red.

The fired colour of a clay depends primarily on its percentage of iron oxide, the state of oxidation of the iron, the degree of subdivision, and the extent of vitrification. Calcium oxide and magnesium oxide are said to "bleach" the colour, and organic matter may lighten the colour by maintaining the iron in the ferrous state.

Vitrification

The manner in which a clay vitrifies is governed largely by the type and amount of fluxing oxides—Na_2O, K_2O, CaO and MgO (see page 26).

Dorset clays and some South Devon clays vitrify appreciably at 1100°C; the North Devon and the remainder of the South Devon clays do not vitrify until a temperature about 1200°C is reached.

Plasticity

"Plasticity" is a vague term, implying several different properties such as "yield value", "water tolerance", binding power and "workability". Ball clays are said to be one of the most plastic clays, being added to poorly-plastic bodies to improve their cohesion. Those containing the most organic matter tend to be the most plastic, which is confirmed by results obtained by Atterberg's method and from the sand method (page 44).

1.5.7 China Clay

China clay is the only important residual clay in the British Isles. Formed from granite rock by hypogenic action, it occurs in the western and central parts of the St. Austell granite, the south-western part of the Dartmoor granite, and the western and southern parts of the Bodmin Moor granite (Figure 26).

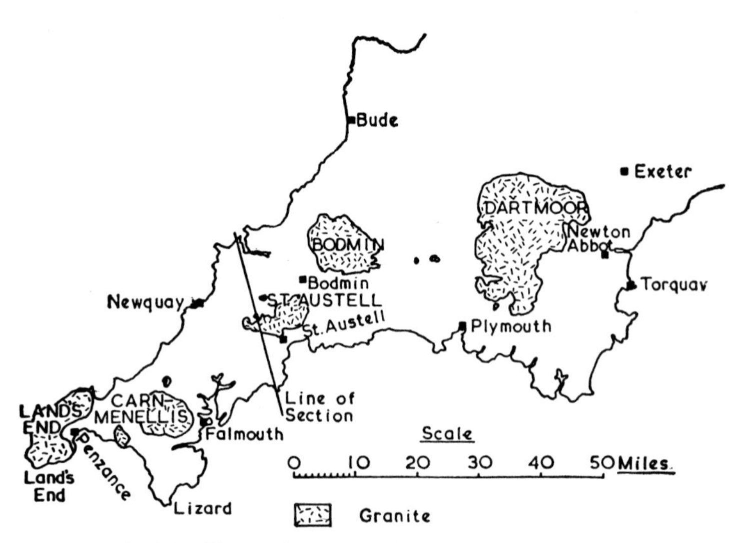

Fig. 26.—The granite masses of south-western England.

The hypogenic agents responsible for the alteration of granite rock were superheated steam and hot acid gases containing compounds of boron and fluorine which, emitted from the interior of the earth, passed upwards through fissures in the granite, decomposing it. The chemical reactions that occurred are not fully understood, but the main action was on the potash feldspar, which broke down to form kaolinite, at the same time releasing silica and potash. The other components of the granite—quartz and mica—were apparently little affected, but small amounts of fluorspar (CaF_2) and tourmaline (a silicate mineral containing fluorine and boron) were also formed.

Extraction

The characteristic way in which this residual clay occurs—as a finely-divided alteration product in granite rock—enables a special method of extraction to be employed. Large pits are dug in the altered granite rock, forming a kind of quarry, and high-pressure jets of water are directed upon the pit walls, removing the fine clay and leaving behind much of the granite. The clay–water suspension so formed is channelled into troughs where it is left for a short time to settle, much of the coarse mica and quartz particles falling to the bottom. The purified suspension is then pumped up to ground level into shallow channels and tanks, where more quartz and mica are removed by sedimentation (Figures 27 and 28).

Fig. 27.—A china clay pit near St. Austell, Cornwall, (photograph by courtesy of English Clays, Lovering Pochin & Co. Ltd.).

Excess water is then removed by a long period of settling (in order to sediment the fine clay particles) or by filter-pressing. Finally the wet clay, at this stage having a putty-like consistency, is dried in special kilns.

Fig. 28.—A china clay works, showing settling tanks (photograph by courtesy of English Clays, Lovering Pochin & Co. Ltd.).

Producers of china clay are now marketing various "standard" grades of china clay, classified according to particle size, wet-to-dry contraction, etc., and they claim that the day-to-day variation of properties in any one grade is kept within narrow limits.

Composition

Owing to the efficiency of the method of extraction, English china clay is one of the purest sources of kaolinite. The percentages of SiO_2 and Al_2O_3 are very close to those of the pure mineral (46.5% and 39.5% respectively), and the alkalis total less than 2%; the iron content (expressed as Fe_2O_3) lies between 0.5% and 1.2%. A calculation of the mineralogical composition indicates approximately 85% or more of kaolinite, with some 12% of "mica". The remainder is mostly quartz and miscellaneous oxides, although recent research indicates the presence of 1% or less of montmorillonite; there is virtually no organic matter.

Particle Size

The particles of china clays do not cover as wide a size range as those of ball clays. In the latter, particles as small as 0.02μ have been found, whereas in china clays particles less than 0.3μ diameter are uncommon.

The following data are quoted for the size distribution of a number of standard grades of china clay (Table 13).

Table 13. Particle-size Distribution of China Clays

Type ot clay	Particle size (%)		
	$<1\mu$	$<2\mu$	$>10\mu$
Porcelain clay	61	73	5
Bone china clay	36	47	13
Sanitary clay	20	31	23
Earthenware clay	30	43	24

Plasticity

Tables 12 and 13 show clearly that china clays on the whole contain less fine material than do ball clays, consequently they are less "plastic" than ball clays, in the sense that they have less binding power. It follows therefore that bodies containing china clay and little or no ball clay are "short" and difficult to manipulate.

Cation Exchange

China clays have a comparatively low cation-exchange capacity, ranging from about 2 to 10 m.e./100 g. As with ball clays, the chief exchangeable ions are H^+, Ca^{2+}, Mg^{2+}, Na^+ and K^+.

Deflocculation

Owing to their low cation-exchange capacity, Ca^{2+} china clays require, in general, less deflocculant than ball clays and are more sensitive to over-deflocculation. Sodium silicate alone does not completely deflocculate; polyphosphates are said to be more effective, but deflocculation with them is not permanent and they are liable to attack plaster moulds.

Wet-to-Dry Shrinkage

Less water is generally required to render china clays workable than is needed for ball clays and on this account the wet-to-dry contraction of china clays is considerably the lower.

Green Strength

The green strength of china clays, when they have been thoroughly dried at 110°C, is much less than that of ball clays, (from 60 to 390 lb/in^2 at 0% moisture), probably because china clays contain less fine material and less organic matter than do ball clays.

Fired Colour

Since china clay contains a very small percentage indeed of iron oxide, it fires whiter than any other clay, and it is much valued on that account.

Firing-shrinkage

Values of from 10 to 13% are quoted for the linear shrinkage of china clays at 1280°C.

Vitrification

China clay, because it contains very little fluxing oxides, shows no appreciable vitrification at 1200°C.

1.5.8 Fireclays

Fireclays are sedimentary clays that were laid down in the Carboniferous Period, most of them being found in the Coal Measures. Although the name "fireclay" suggests a clay that can withstand heat (i.e. a refractory clay) a large proportion of the so-called fireclays are not very refractory, but are used for making sanitary fireclay, buff tiles, engineering bricks, etc.

These deposits have been subjected to considerable pressure, because of the depths to which they have been submerged, and often they are very hard and compacted; some have been so altered by high temperature and pressure that they have formed *shales* or *slates.*

Vegetable matter, deposited on top of successive clay beds, decayed and was transformed into bituminous coal and so we now find alternate seams of coal and clay. Generally speaking, the Coal Measure clays can be divided into two groups : (1) underclays, and (2) shales. The underclays are situated immediately beneath the coal seams and, unlike the shales, they are not laminated, and they are not so hard. Shales form the bulk of the Coal Measure clays, and depending on their composition, are termed calcareous (containing $CaCO_3$), sandy, bituminous, oil or alum shales.

Fireclays are found in the following areas of the British Isles :

NORTHUMBERLAND AND DURHAM

Almost all the clays in these counties are extracted as by-products of coal mines or open-cast sites. Al_2O_3 contents vary from 28–36%, the Durham fireclays being slightly higher in Al_2O_3. These clays are used for making sanitaryware or refractories, according to the Al_2O_3 content.

CUMBERLAND

Fireclay deposits lie chiefly along the Coast and under the sea, extending from South of Whitehaven to Wigton. The only important workings are at Lowca, 5 miles south of Workington. Here clay from the Micklam Seam containing about 35% Al_2O_3 is mined and used for making firebricks and casting-pit hollow ware for steelworks. A similar fireclay occurs at St. Helens colliery, just north of Workington, but this is not at present being utilized.

LANCASHIRE and part of CHESHIRE

The fireclays cover a triangular area corresponding to the coalfield, running from Colne in the north to Huyton in the south-west and to Stockport in the south-east. In addition, there is a narrow area extending to Macclesfield and to Chapel-en-le-Frith.

One of the most useful fireclays here is the underclay of the Mountain Mine coal, worked in the Blackburn-Burnley district. Other useful underclays are worked in the Wigan, Horwich and St. Helens district, also near Stockport, Macclesfield, Oldham and Bolton.

NORTH STAFFORDSHIRE

The fireclays again correspond with the North Staffordshire coalfield, extending from Congleton in the north to Longton in the south, and bounded in the east and west at Oakamoor and Madeley respectively. Many of these clays are not highly refractory but some are used for making bricks and tiles.

SOUTH STAFFORDSHIRE AND WORCESTERSHIRE

The coalfield extends from Armitage in the North to Halesowen in the South, with Wolverhampton and Walsall on the Western and Eastern boundaries.

The best known fireclays of this district are those of Stourbridge. The Old Mine clay, with an alumina content varying from 25–40%, was of much value as a refractory. This is now worked out but has been replaced by deeper seams called the New Mine clay with some 30–40% Al_2O_3, of which large reserves are available. The clays are used for making firebricks, sanitaryware and salt-glazed pipes, according to the Al_2O_3 content.

WARWICKSHIRE

The most important fireclay-producing area is between Tamworth and Nuneaton (the so-called Nuneaton clay). The latter is not very refractory but is used for making building-bricks, salt-glazed pipes, tiles and chemical stoneware.

DERBYSHIRE AND LEICESTERSHIRE

The fireclays fall into two main groups : (1) near Swadlincote, Church Gresley and Woodville, and (2) east of the Erewash valley. These clays are less refractory than those of Stourbridge, since they contain some 4% of lime, magnesia, and alkali oxides.

YORKSHIRE

Fireclays of widely varying composition occur, and are concentrated in the Leeds and Halifax districts. Many of them provide high-quality refractory bricks; others, which are less refractory, are used for making salt-glazed ware and drain-pipes. Somewhat less refractory fireclays of a more siliceous nature are obtained from the Huddersfield and Sheffield districts.

LINCOLNSHIRE

Fireclay is confined chiefly to Stamford and district.

SHROPSHIRE

The coalfield extends from Dorrington, almost due south to Bewdley and Stockton (Worcs.) but the chief fireclay area within this is in the Coalbrookdale district. These fireclays are contaminated with nodules of ferrous carbonate, but are used for making wall- and floor-tiles.

SCOTLAND

The fireclays, with the Coal Measures, are concentrated in the Forth-Clyde Valley. Some of the best refractory clays come from this area, notably the Glenboig and Bonnybridge fireclays, which have an Al_2O_3 content of nearly 40%. The upper fireclay seams in Ayrshire are exceptional in that they contain free Al_2O_3 in the form of bauxite, which naturally gives them a high refractoriness. All the fireclays just mentioned can be made into high-grade refractory products.

Slightly less refractory clays, like those worked near Barrhead, are used for making ordinary firebricks and sanitaryware. Fireclays also occur in various other parts of Scotland, but they are less refractory and serve the local markets only.

NORTH WALES

Fireclays are worked at Mold, Ewloe, Buckley, Ruabon and Wrexham.

SOUTH WALES

The main fireclay workings are in the vicinity of Neath, Swansea, Llanelly, Merthyr Tydfil and Pontypridd. Most of the Welsh fireclays are highly siliceous and of only moderate refractoriness.

IRELAND

Some fireclay deposits are worked in Co. Tyrone; they are mostly of inferior grade and unsuitable for refractory products, but some are said to be very suitable for the manufacture of salt-glazed pipes.

Extraction

Fireclays are extracted by open-pit methods, as described for brick-clays. If the seam is a deep one, and sufficiently valuable, it may be extracted by deep mining, as for coal, particularly if the coal can be extracted at the same time.

Composition

Some idea of the composition can be obtained from Table 14, which shows the ultimate analyses of a number of fireclays used in the sanitary and tile industries. A wide range of compositions is shown; for instance, for a total of 126 samples, the SiO_2 contents ranged from 44.5 to 81.4%, and Al_2O_3 contents from 11.9 to 37.8%. It should be emphasized that this survey did not, of course, include the more refractory clays used for the making of firebricks, some of which have Al_2O_3 contents of some 40%. The highest proportion of carbon (determined by combustion) in the whole series was 5.7%—much lower than was found for ball clays and brick-clays. Carbon is of interest because of the effect it may have on certain physical properties (see below) and on firing.

Rational Analysis

The chief minerals in the fireclays are said to be a poorly-crystallized kaolinite, mica, and quartz. Minor impurities such as organic matter, carbonates of calcium, magnesium and iron, pyrites, hydrated iron oxide and anatase (a form of TiO_2) are also frequently present.

The percentages of the principal minerals can be calculated by the method as described for ball clays, but is subject to the same errors (Table 7). In particular, the poorly-crystallized kaolinite departs somewhat from the "ideal" formula, some Fe replacing Al; in addition, the micas are believed to differ appreciably from normal muscovite.

The underclays are said to contain less impurity than the shales (Table 14).

Particle-size Distribution

The average particle-size distributions for a number of fireclays are shown in Table 14. Particles larger than 2μ equivalent spherical radius are determined by a method based on settling under gravity, but for smaller particles a centrifugal method has to be employed. It is unfortunate that more information on the distribution below 2μ is not available, since it is this fraction that contributes most to the specific surface area and related properties. Nevertheless, the data available does correlate to some extent with other physical properties.

It is also interesting that, on the whole, the clays from the north of the British Isles are coarser than those from further south. Since the former were deposited by running water which is believed to have travelled from north to south, it is to be expected that the coarser particles would be deposited first in the north.

Critical Moisture Content

Since the critical moisture content (Table 14) is a measure of the water remaining in the pores of a clay article at the point at which shrinkage ceases, it ought clearly to be related to the degree of packing of the particles and therefore to the size distribution. Although there is some suggestion from tests carried out that this is so, the correlation is not good, and other factors are obviously involved. One of these factors is undoubtedly the degree of flocculation, since it markedy affects the effective aggregate size, and also the thickness of the lyosphere attached to the colloidal particles.

Dry Strength

The dry strength of a number of fireclays has been determined (Table 14). These values were for test-pieces that have been dried at 110°C; this procedure is adopted because the presence of quite small amounts of moisture (less than 1%) has a marked effect on the results.

Again it is difficult to relate the dry strength to any other physical characteristic, but there is some suggestion that it depends chiefly on size distribution. However, it has been found that the exchangeable cations also have a considerable effect, as discussed on page 42.

Deflocculation

The response of various clays to deflocculants, which has been discussed under ball clays, in general also applies to fireclays. The most commonly-used deflocculants in the sanitary earthenware industry are mixtures of sodium carbonate and sodium silicate in various proportions. It has been found that the casting of bodies containing fireclay can be well controlled by adjusting the flow properties, the amount of deflocculant and the slip density. Difficulties have been experienced with H^+ clays which can only be deflocculated with sodium silicate and not with the carbonate-silicate mixtures.

Nevertheless, in the determination of size distribution, the deflocculant used is a 1:1 mixture of NaOH and "Calgon" (sodium hexametaphosphate); whether it is more effective than sodium carbonate–silicate mixtures is not yet established.

Refractoriness

The refractoriness of a material is a measure of its ability to withstand a high temperature without loss of shape, and is determined by heating in a furnace a small cone, made from the material, side by side with other standard cones of known refractoriness, (called Seger Cones) until the cone being tested begins to fuse, indicated by its bending over. The test cone and the standards are removed from the furnace, and compared. The test cone has the same refractoriness as that of the standard cone that has slumped to the same extent, and can be expressed as a cone number. Provided that the test is carried out under strictly controlled conditions, the refractoriness can be expressed as a temperature, by reference to a chart of standard cones. A clay suitable for the manufacture of firebricks should have a refractoriness of cone 30, corresponding to a temperature, under specified conditions, of 1670°C.

As already mentioned, the data available do not include the most refractory fireclays, which are found particularly in Scotland, but from the values in Table 14 it is interesting that the Scottish fireclays cover the greatest range of refractoriness, and include the highest value found in this group—over 1770°C (over cone 35). The Staffordshire and Shropshire fireclays, used for making glazed tiles, have the lowest mean refractoriness of the five main groups.

Since so many factors are involved, it is difficult to correlate refractoriness with chemical composition but, as a general rule, the most refractory fireclays are those that contain a high percentage of Al_2O_3 and a low percentage of alkalies (Na_2O and K_2O). The most refractory fireclays contain 40% or more Al_2O_3.

Firing-shrinkage

Clays in general shrink when fired because the pore spaces gradually become filled with molten material which eventually, on cooling, solidifies to form a glass. If the firing is sufficiently prolonged, all the pore spaces become filled and the body is then said to be vitreous, and has no porosity.

Other non-clay materials can affect the shrinkage very markedly, however. Prefired material, known as "grog," is added to fireclay for the purpose of reducing the shrinkage of the resulting firebricks. Since when quartz is converted to tridymite or cristobalite a considerable expansion occurs, a high proportion of quartz in a clay will result in a low firing-contraction. High contents of mica tend to give a high contraction, probably because the associated alkalis produce a high proportion of melt at the firing-temperature.

Since the reactions that occur during firing never go to completion, the size distribution of the clay must also influence firing-shrinkage. Very fusible clays often evolve gases at the temperature

Table 14. Some Properties of Fireclays from various Localities

Chemical Analysis (%)	*South Midlands*		*Yorkshire*		*Northumberland and Durham*		*Scotland*		*Staffordshire and Shropshire*	
	Mean	*Range*	*Mean*	*Range*	*Mean*	*Range*	*Mean*	*Range*	*Mean*	*Range*
SiO_2	56.5	45.9—67.3	60.1	53.4—76.5	59.5	48.6—81.6	57.4	44.5—79.6	60.9	44.5—71.1
Al_2O_3	26.6	19.4—33.2	25.7	15.5—30.0	26.0	12.2—32.5	26.4	11.9—37.8	23.7	16.7—33.8
Fe_2O_3	2.4	1.3— 5.6	1.7	0.8— 2.8	1.6	0.6— 3.4	1.9	0.7— 4.2	2.2	1.1— 4.4
TiO_2	1.5	1.3— 2.1	1.2	1.0— 1.6	1.2	1.0— 1.6	1.2	0.7— 1.9	1.3	1.1— 1.5
CaO	0.3	0.1— 0.8	0.4	0.2— 0.5	0.4	0.1— 0.8	0.3	0.1— 0.6	0.3	0.0— 1.6
MgO	0.5	0.2— 0.7	0.5	0.2— 0.8	0.5	0.2— 0.8	0.5	0.2— 1.0	0.6	0.0— 1.7
Na_2O	0.2	0.1— 0.5	0.15	0.1— 0.3	0.1	0.0— 0.3	0.1	0.0— 0.4	0.2	0.0— 0.4
K_2O	1.4	0.4— 2.6	1.3	0.7— 2.7	1.7	0.9— 2.6	1.2	0.2— 1.7	1.8	0.8— 3.5
Ignition loss	10.6	6.8—17.1	9.3	5.3—12.6	9.0	3.9—13.1	10.9	4.4—14.8	8.9	5.6—19.6
SO_3	0.4	0.1— 1.0	0.5	0.2— 1.1	0.2	0.1— 0.4	0.3	0.1— 0.4	0.3	0.1— 1.8
Carbon	1.6	0.1— 5.7	0.9	0.3— 2.1	0.9	0.3— 2.2	1.7	0.3— 4.8	1.2	0.2— 5.6
Kaolinite	53.3	35.6—75.6	52.6	28.7—68.4	50.4	23.5—67.8	55.5	23.4—93.3	42.9	26.9—67.6
Mica	14.8	6.5—25.8	12.6	6.5—27.8	15.7	7.6—24.0	12.1	2.2—28.9	17.7	5.8—30.6
Quartz	25.4	9.0—44.6	30.0	19.0—58.0	29.3	10.4—67.0	26.2	0.1—65.6	33.1	4.7—50.8
Carbonaceous matter	2.2	0.2— 7.7	1.3	0.0— 3.3	1.2	0.0— 3.0	2.6	0.5— 6.1	2.0	0.3— 9.9
Particle size (%)										
μ										
<0.1	9	4 —12	4	3 —6	3	2 — 7	3	2.5— 7	—	—
0.1— 2.0	38	9 —47	24	17 —30	22	9 —30	29	13 —53	49.8	33.0—83.5
2.0— 5.0	19	10 —26	13	8 —20	13	8 —18	14	3 —23	22	8 —35
5.0—10.0	9	1 —15	10	6 —28	11	5 —18	9	0 —19	12	2 —21
10.0—25.0	10	5 —17	12	8 —17	14	4 —20	10	1 —16	10	0 —31
Critical moisture content (%)	11.0	9.0—15.1	10.1	9.1—11.5	9.6	8.5—11.4	10.0	8.5—12.1	11.5	9.1—15.3
Modulus of rupture (unfired) (lb/in²)	314	155— 540	304	170— 480	266	190— 500	318	180— 950	430	220— 840
Refractoriness (°C.)	1647	1530—1710	1630	1460—1680	1662	1595—1700	1660	1490—>1770	1595	1410—1720
Linear firing-shrinkage (%)	4.6	2.0— 6.0	2.8	0.0— 4.0	2.8	0.0— 5.0	3.2	0.5— 5.0	4.2	0.9— 8.5

of firing, causing local expansion known as "bloating," which will obviously reduce the overall contraction.

The wide range of firing-shrinkages is shown by the values given in Table 14.

Vitrification

As will be realized from the above discussion, the vitrification of a clay is closely related to fired porosity and to firing-contraction. Of the fireclays examined, none of the South Midlands, Yorkshire, Northumberland, Durham and Scottish groups had vitrified at 1200°C, the average porosity after firing at the latter temperature being some 20%. On the other hand, of the Staffordshire and Shropshire group, six samples out of a total of 64 had completely vitrified at 1200°C, which is consistent with their lower average refractoriness.

Fired Colour

Little need be said about the fired colour of fireclays. Because of the appreciable amount of iron oxide they contain, they fire buff to reddish brown in an oxidising atmosphere and so are not suitable for use in the production of whiteware.

Plasticity

Although the binding power of fireclays is less than that of ball clay, owing to their greater proportion of coarse material, it is generally adequate for the making of refractory bricks and other articles, even when grog has been added. Unlike the organic matter of ball clays, that of fireclays does not appear to contribute to plasticity. Underclays, which contain less free silica and other impurities, tend to be more plastic than shales.

Since plasticity is fully developed only when all the available surface of the fine particles is accessible to water, the plasticity of fireclays is sometimes enhanced by weathering, which breaks up the hard, compacted grains.

1.5.9 Brick Clays

Occurrence

Since brick-clays constitute a great variety of sedimentary deposits, ranging from the Devonian to the Recent Period of geological time, and are very widespread, it is convenient to classify them according to the geological formations in which they occur (Table 9).

DEVONIAN

Clay deposits of the Devonian Period are used for brick-making in South Devon and Cornwall and around Cardiff and Newport.

CARBONIFEROUS

Non-refractory clays of the Carboniferous Period, particularly those of the Coal Measures, are used to a considerable extent by the brickmaking industry. Suitable brick-clays of Carboniferous age are worked in the Glasgow district, and in the neighbourhoods of Durham and Newcastle-on-Tyne. In North Staffordshire, clays of the Upper and Lower Coal Measures, including the so-called Etruria Marls, are used for the manufacture of bricks and roofing-tiles. Other Coal Measure clays utilized for brick-making occur in Leeds, Halifax, Huddersfield and Wakefield (Yorkshire), Mansfield (Nottinghamshire), Ashby-de-la-Zouch and Measham (Leicestershire), Nuneaton (Warwickshire), Tipton, Bilston, Willenhall and Walsall (South Staffordshire), Bristol (Gloucestershire), and in Cardiff and Ruabon (Wales).

THE KEUPER MARLS

This important group of clays of the Triassic Period (Table 9), extends across the country from Sidmouth (Devon) to the mouth of the Tees, and is utilized for brick-making around Birmingham, Mapperley (Nottinghamshire) and in Leicester. Their use is often limited because of contamination with rocksalt and gypsum ($CaSO_4.2H_2O$).

THE OXFORD CLAY

A Jurassic deposit rich in organic matter that is worked at Bletchley, (Buckinghamshire), Kempston Hardwick (Bedfordshire), Leighton Buzzard (Bedfordshire), Stewartby (Bedfordshire) and Peterborough (Northamptonshire). Some 40% of the total brick production of the U.K. is made from this clay.

THE HASTINGS BEDS

Bricks are made at Hastings and Bexhill from these Lower Cretaceous clays.

THE WEALD CLAY

Also of the Lower Cretaceous Period, the Weald Clay occurs in Sussex, Isle of Purbeck (Dorset), and in parts of the Isle of Wight, and is worked in all these areas.

THE GAULT CLAY

Belonging to the Upper Cretaceous System, the Gault forms a thin, straggling line across Dorset and Wiltshire, running from Shaftesbury (Wiltshire) to north of Abbotsbury (Dorset); it is located again at Westbury, can be traced as far as Devizes (Wiltshire), and outcrops again in east Hampshire, Berkshire, Oxfordshire, Buckinghamshire, Bedfordshire and Cambridgeshire. Part of it also branches across Surrey, Sussex and Kent. Near Eastbourne it is used for the manufacture of bricks, drain-pipes and tiles, and at Shaftesbury and Devizes for brick-making.

EOCENE DEPOSITS

Clays of Eocene age occur in two principal areas—the Hampshire Basin and the London Basin. These deposits are used for brick-making in various localities, and include the clays of the Reading Beds, the London Clay, the Bracklesham Beds (Isle of Wight) and the Barton clay (Hampshire).

OLIGOCENE DEPOSITS

These deposits occur only in the Hampshire Basin and in Devonshire. Among those used are part of the Headon Beds near Bournemouth and the Hamstead Beds (Isle of Wight).

PLIOCENE DEPOSITS

Deposits of this system are worked in a few places in East Anglia (Norfolk and Suffolk). Many are of no use to the brick-maker, because they are too calcareous and fusible.

PLEISTOCENE DEPOSITS

These deposits include the alluvium and brick-earths; they are generally found in the river valleys, for example in the Thames Valley, North Kent, and the Forth, Clyde and Tay estuaries. Many former brick-earth deposits have now been worked out.

Boulder Clay is of glacial origin, belonging to the Holocene or Recent Period and, as the name suggests, is associated with boulders and small stones. Deposits of Boulder Clay are scattered over East Anglia, Leicestershire, Staffordshire, Shropshire, Warwickshire,

Table 15. Analyses of Bricks made from Clays of various Geological Deposits (%)

	Alluvial	*London stock*	*Glacial*	*Oligocene*	*London clay*	*Gault*	*Weald*	*Oxford Clay (Fletton)*	*Middle Lias*	*Keuper Marl (Upper)*	*Keuper Marl (Middle)*	*Keuper Marl (Lower)*	*Permian*	*Etruria Marl*	*Coal Measure*	*Coal Measure*	*Coal Measure*	*Devonian*
SiO_2	64.7	68.7	62.5	77.8	64.4	47.2	68.4	56.2	57.8	58.3	46.2	66.0	60.1	62.7	54.9	61.7	61.9	59.6
Al_2O_3	12.7	11.0	18.6	15.8	15.8	19.4	17.2	20.9	23.2	15.3	13.7	13.9	16.7	23.1	34.9	21.6	24.0	19.9
Fe_2O_3	8.3	7.0	6.6	0.8	7.9	6.1	6.3	6.0	9.3	6.0	6.0	6.8	5.8	8.4	3.4	8.0	8.7	11.4
TiO_2	1.6	0.7	0.9	2.2	1.6	0.8	1.3	0.5	1.2	0.7	0.8	0.6	0.5	1.2	0.6	1.2	1.1	1.2
CaO	7.9	8.1	4.1	0.3	1.1	19.2	1.9	8.1	1.0	6.2	11.4	3.5	6.5	0.9	2.1	0.6	0.6	0.2
MgO	1.9	0.8	3.4	0.4	2.4	1.9	1.2	1.7	2.5	7.3	12.8	2.9	4.2	1.2	0.7	1.0	1.1	1.2
Na_2O	0.4	0.8	0.5	0.5	0.5	0.6	0.5	0.5	0.9	0.7	0.3	0.6	1.2	0.4	0.1	1.2	0.2	1.0
K_2O	1.5	2.0	2.9	1.8	3.2	3.1	2.2	3.6	2.9	4.7	3.3	4.1	3.3	2.6	2.6	3.1	1.6	4.2
SO_3	1.4	0.6	0.4	0.5	2.3	1.4	0.7	1.9	0.3	0.5	5.6	1.0	0.9	0.7	0.1	1.5	—	0.1
Loss	0.3	nil	0.3	0.3	1.1	0.4	0.3	0.6	1.0	0.4	0.5	0.2	0.3	nil	0.6	0.1	0.5	0.5

Wales, the North of England and the South of Scotland. Works making bricks from Boulder Clay are located at Buckley (Flintshire), around Manchester, Peel (Isle of Man) and at Sudbury (Suffolk).

Composition of Brick clays

The principal minerals in brick-clays are kaolinite and chlorite, with illite, quartz, and organic matter; many brick-clays contain considerable amounts of iron oxide and calcium carbonate. Because the composition of the minerals, particularly chlorite, is uncertain, it is difficult to calculate a "rational analysis" for brick-clays, as has been done for ball clays.

From the chemical compositions of bricks made from various clays (Table 15) it can be seen that although the range is wide, most of the clays contain considerable proportions of iron oxide, and some have a good deal of calcium oxide (reported as CaO in analyses, but probably present in the natural clay as $CaCO_3$). Soluble salts, chiefly calcium sulphate, sometimes occur in these clays and can cause efflorescence in bricks made from them.

Physical Properties

Some of the more important properties of some brick-clays are given in Table 16.

Particle-size Distribution

The size distribution of particles in brick-clays ranges from 2.5 mm. down to 2μ but, as pointed out under ball clays, it is the distribution of particles below 2μ that is the most important. The values in Table 16 refer to the latter size, which is usually said to consist entirely of clay; this is not strictly true, because some non-clay impurities as small as this do exist, but identification of the "less than 2μ range" with "clay" is a fair approximation.

The value given for the percentage less than 2μ is the average of between 50 and 100 clays of one particular formation.

Deflocculation

The response of a brick-clay to deflocculants is of little technological interest, since such clays are not cast; deflocculants are used, however, in determining the size distribution by sedimentation methods, and it is of interest that, for the Etruria Marls, a mixture of 5 parts by weight of "Calgon" to 1 part of NaOH was a more effective deflocculant than sodium oxalate. All the other clays mentioned were deflocculated well with this mixture. This behaviour is probably related to the exchangeable ions of the clays.

Working-moisture Content

The moisture content at which a clay is workable for some particular process depends on the proportion of clay mineral present, the surface area of the clay, and on the type and amounts of the exchangeable ions. Since the compositions and particle-size distributions of the brick-clays cover a wide range, it is reasonable to expect a wide range of working moisture contents. However, even where the size distributions appear to be similar, as in the Etruria Marls, the Weald Clays and the Boulder Clays, the working moisture contents differ considerably. This may be partly because the distributions below 2μ (the important range) are different, or because the exchangeable ions are different.

Firing-shrinkage

To obtain a valid comparison of the firing-shrinkages of different clays, they would have to be fired at the same temperature; this is impossible in practice, because the fusibility varies so widely that some clays would vitrify at a temperature too low to cause appreciable shrinkage in others.

For this reason the firing-shrinkages in Table 16 were determined at temperatures suitable to the particular type of clay, and therefore vary. In particular, the Boulder Clays were calcareous (contained much $CaCO_3$) and fusible, and so could not be fired at such a high temperature as some others.

Sometimes gases are evolved when a clay is heated to a very high temperature and cause bloating; clearly such clays are likely to expand rather than to contract on being fired, which often makes correlation of the firing-shrinkages with other properties difficult.

Since the shrinkage during firing is caused by reactions associated with the clay mineral, the percentage shrinkage at a given temperature can be correlated approximately with the percentage of material smaller than 2μ.

Fired Colour

The fired colour of a clay is due principally to its content of iron oxide, but also depends on the firing-atmosphere (whether

Table 16. Physical Properties of some Brick-clays

Clay	*Particle size (% smaller than 2μ diameter)*		*Working moisture content (% dry basis)*		*Linear firing shrinkage (%)*		*Fired colour for different temperatures*	
	Range	*Average*	*Range*	*Average*	*Temperature (°C.)*	*Range*	*Temperature (°C.)*	*Colour obtained*
Coal Measure shale (outcrop)	16—78	34	15.1—31.2	19.5	1050° 1180°	0.2— 7.7 2.0— 9.7	1050° 1180°	Light salmon, red pink, light cream. Green-grey, dark brown, chocolate brown, greenish-yellow.
Coal Measure shale (from coal seams)	14—43	26	15.0—25.1	17.8	1000° 1180°	0.4— 5.1 0.7— 7.7	1000° 1180°	Salmon pink, pale pink, buff, white. Green-grey, light green-grey, cream.
Etruria Marls	24—74	47	15.2—30.2	23.6	1180°	0.6—12.1	1000° 1180°	Light buff, pink buff, light red, light brown, light-chocolate brown, pink cream. Dull brown, red brown, purple brown, medium red, stone, greyish buff.
Weald clay	15—85	47	23.8—42.2	30.9	900° 1100°	0 — 3.4 2.5— 8.6	850° 1200°	Salmon pink, light red-brown, buff. Dark red-brown.
Boulder clays	30—60	47	17.1—39.6	28.8	900° 1070°	0.1— 4.3 2.5— 9.4	880° 1070°	Salmon pink, light brown. Light chocolate-brown, light brown.

oxidizing or reducing) and the temperature of firing, and is modified by other constituents such as Al_2O_3, CaO, and MgO. Under oxidising conditions, iron oxide fires buff to brown, but the latter is "bleached" by the constituents mentioned above. On the whole, the colour is darkened by increasing the firing-temperature (Table 16).

Vitrification

Since the oxides of sodium, potassium, calcium and magnesium act as fluxes and promote vitrification, brick clays containing a good deal of $CaCO_3$ are readily fusible, and have to be fired at a comparatively low temperature; for building-bricks the range of firing-temperatures is from 950° to 1200°C.

Extraction

Brick-clays are mostly extracted in open workings, but if they outcrop on the side of a hill, with an extensive overburden, the most convenient method is to tunnel into the side of the hill at the level of the seam (drift or tunnel-mining). If the seam is not more than 20 ft. below the surface, the overburden is stripped off by means of skimmers, power shovels, bulldozers or shovel-dozers, and the underlying clay is removed with a dragline scraper or skimmer (Figure 29). Sometimes shallow pits are dug out and the clay is stripped from the sides of the pit with a bucket-type mechanical excavator (Figure 30).

Stoneware Clays. These clays include various readily-fusible plastic clays that are not white-burning. Some non-refractory clays of the Coal Measures are used as stoneware clays.

Pipe-clays. Plastic, fusible clays, deposits of which are found in various geological formations.

Boulder Clays. Of glacial origin, Boulder Clays occur in the Recent or Holocene Period. (See p. 76).

Bentonite. A general term for a montmorillonite clay. It is not common in the British Isles, but is found in considerable quantity in the U.S.A. and elsewhere. Bentonites are exceedingly plastic, with a high cation-exchange capacity (about 120 m.e./100g), the chief exchangeable cation being Na. Bentonite is often added to bodies to improve their plasticity.

Fig. 29.—A scraper removing overburden, (photograph by courtesy of T. G. Carruthers, The University of Leeds).

Fig. 30.—A bucket-type mechanical excavator removing brick clay, (Geological Survey photograph A.8973, reproduced by permission of the Controller, H.M. Stationery Office).

Fuller's Earth. An impure montmorillonite clay, contaminated with quartz, calcite, and other minerals; it is one of the few sources of montmorillonite in this country. Fuller's Earth was formerly used for "fulling"—removing fatty substances from wool, this property being due to its large surface area, which gives it great adsorptive power. It has a high cation-exchange capacity, the chief exchangeable cation being Ca. The only important deposit of Fuller's Earth is near Nutfield, Surrey.

1.6 OTHER SILICATES

1.6.1 Sillimanite, Kyanite, Andalusite

These three minerals all have the empirical formula Al_2SiO_5 (or $Al_2O_3.SiO_2$), and contain some 63% of Al_2O_3. Sillimanite comes chiefly from South Africa and India, kyanite from the U.S.A. and India, and andalusite from South Africa and the U.S.A.

Structure

In all three minerals there are parallel chains of Al-O groups, which are linked sideways by Si and Al ions alternately; the three structures are similar and differ only in detail. A full discussion of the differences between the three structures is beyond the scope of this work; the chief point of difference is in the co-ordination of the Al^{3+} ions.

Properties

Sillimanite, kyanite, and andalusite are non-plastic and so are processed by fine grinding, with the addition of a binder or plasticizer. When heated at about 1550°C, all three minerals decompose to form *mullite* and cristobalite :

$$3Al_2SiO_5 \xrightarrow{1550°C} \underset{\text{Mullite}}{Al_6Si_2O_{13}} + SiO_2$$

(Mullite is discussed in the next section).

Consequently, when bricks are made from any of the above three minerals, the final product is the same, since during firing the reaction shown takes place.

The sillimanite group of minerals has a high refractoriness and is very resistant to attack by alkaline slags.

1.6.2 Mullite

Occurrence and Composition

Natural mullite is not very common; it is named after one of the few known deposits, on the Isle of Mull, and has the approximate formula $3Al_2O_3.2SiO_2$, or $Al_6Si_2O_{13}$, and so contains 72% of Al_2O_3.

Synthetic mullite is commonly used, and can be made by heating a mixture of pure Al_2O_3 or bauxite with clay or sillimanite. Mullite is a common constituent of fired pottery bodies and refractories and, under the microscope, appears as long prism-shaped crystals of nearly square cross-section.

Structure

Mullite has a structure similar to that of sillimanite, in which some silicon has been replaced by aluminium. Aluminium can also be taken up into solid solution in variable proportions, so that the chemical composition may deviate slightly from that quoted above.

Properties

Mullite is very refractory and resists alkaline slags. It dissociates at 1810°C to give corundum and a siliceous liquid.

1.6.3 Steatite

Steatite is mineralogically the same as talc, also known as soapstone or French chalk; the term steatite, however, is usually reserved for the massive form.

Talc was formed by the hydration of magnesium-bearing rocks under pressure, and may be derived from basic igneous rocks or from dolomite or marble. It occurs in many parts of the United States, in Germany, France, Egypt, Morocco and Indo-China.

Structure

Natural talc is not pure, but contains Al, Ca and Fe. As already described (see section on montmorillonites) talc is a hydrated aluminium silicate having the formula $Mg_3Si_4O_{10}(OH)_2$. Structurally it is composed of a layer of brucite, $Mg(OH)_2$, "sandwiched" between two silica-type layers (see section on montmorillonites).

Properties

Like the montmorillonites, talc cleaves easily along one plane; it is soft and has a characteristic "soapy" feel. In the massive "block" form it can be readily machined, and fired to produce a strong body. Unlike the montmorillonites it is not plastic and so clay is sometimes added as a binder.

When heated about 900°C, talc is decomposed with the elimination of the combined water. At still higher temperatures—about 1300°C—recombination occurs and clinoenstatite, $MgSiO_3$, is formed, and the reaction can be written:

$$Mg_3Si_4O_{10}(OH)_2 \longrightarrow \underset{\text{Clino-enstatite}}{3MgSiO_3} + SiO_2 + H_2O$$

Uses

Talc, compounded with a little clay and some flux, is used in making *steatite bodies* for low-loss electrical insulation.

With the addition of some 50% clay and some grog, talc is also used in the manufacture of *cordierite bodies.* Their main constituent, cordierite, is a magnesium aluminium silicate of formula $2MgO.2Al_2O_3.5SiO_2$; it is a good electrical insulator at high temperatures, with a very low coefficient of thermal expansion and high thermal shock resistance.

By replacing the feldspar of an earthenware body with talc, a body of low moisture expansion, suitable for wall tiles, is obtained. For all these purposes, the talc should contain a minimum of iron oxides and alkalis.

1.7 FLUXES

1.7.1 Definition

A flux is a substance that is added to a material to enable it to fuse more readily. In ceramics, and in the pottery industry in particular, fluxes are incorporated in the body in order to lower the temperature at which liquid forms during firing. This liquid, when cooled, forms a glass, which binds the grains of the body together. By means of fluxes, strong articles of pottery or porcelain can be produced by being fired at 1100°—1300°C. It is not desirable to produce more glass than is necessary to give the fired article the required strength, and so the amount of flux is kept within limits.

Although any material that promotes fusion is a flux, for silica-containing materials the most effective fluxes are those that contain the alkali oxides Na_2O, K_2O, or Li_2O. Lime (CaO) and magnesia (MgO) can also act as fluxes, though they are usually in a combined form. While for instance, sodium carbonate, calcium carbonate or borax could be used as fluxes, naturally-occurring minerals that contain Na_2O or K_2O are both more convenient and more economical.

In discussing the fluxes that are used in the ceramic industry, it is important to distinguish between rocks and minerals. A rock is any formation of the earth's crust, and may contain many different minerals, whereas a mineral is a pure, naturally-occurring substance with a definite composition and well-defined physical properties. Most minerals are crystalline and definite chemical formulae can be assigned to them.

1.7.2 Soda- and Potash-bearing Minerals

The more important of these minerals are compounds of Na_2O or K_2O with SiO_2 or with both SiO_2 and Al_2O_3.

The Feldspars

The feldspars are a group of minerals that all have similar chemical formulae. For the ceramic industry the most important are :

Potash feldspar (orthoclase) :	$KAlSi_3O_8$
Soda feldspar (albite) :	$NaAlSi_3O_8$
Lime feldspar (anorthite) :	$CaAl_2Si_2O_8$

Each may be considered as consisting of unit formulae of silica, Si_4O_8, in which one or two of the four Si atoms has been replaced by Al atoms and the valency deficiency made up by the addition of one atom of Na^+, K^+, or Ca^{2+}.

The arrangement of the atoms is somewhat different from that of silica, however. The Si and O atoms are linked so as to form four-membered rings, each ring containing four O atoms; in orthoclase and albite the rings also contain three Si atoms and one Al atom whereas in anorthite, they contain two Si atoms and two Al atoms. The four-membered rings are linked with other rings to form chains which are cross-linked with similar chains via Si–O–Si groups, forming a three-dimensional framework. The Na, K, or Ca atoms, which are large, are situated in large cavities within the framework. Part of a feldspar chain is shown in Figure 31. The Si (and Al) atoms are as usual in four-fold co-ordination with oxygen, but the comparatively large Na, K and Ca atoms are in eight-fold co-ordination.

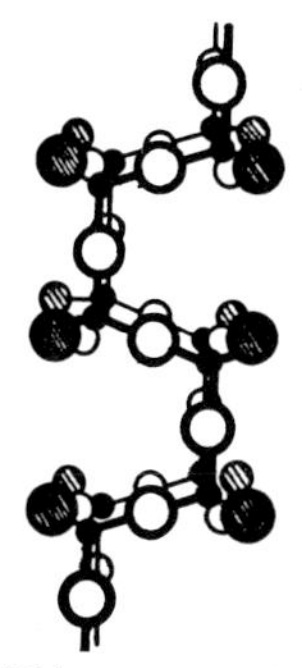

Fig. 31.—The structure of the feldspars : part of a feldspar chain (reproduced from "Structural Inorganic Chemistry", by kind permission of A. F. Wells and the Clarendon Press, Oxford).

The chemical composition of a typical feldspar is shown in Table 18. The iron content is very low and it is therefore very suitable for whiteware.

Of the fluxes used in the ceramic industry the feldspars are the most important but one or two other minerals are sometimes used and so deserve mention.

Nepheline, $Na(AlSi)O_4$, has a structure very similar to that of tridymite, the positions of the Si and O atoms corresponding closely. Nepheline may, in fact, be regarded as two-unit formulae of tridymite, Si_2O_4, in which one of the Si atoms is replaced by Al, and the deficiency in valency is made up by the addition of Na. The Na atoms fit into the large "holes" in the tridymite framework (see Figure 7A).

Like tridymite, nephelite changes to a cristobalite type of structure at high temperatures.

Lepidolite is a lithium-bearing mineral, the constitution of which is represented by the idealized formula $K(AlLi_2)Si_4O_{10}(OH)_2$. It is a mica in which substitution of Li^+ for Al^{3+} in the ocatahedral layer is balanced by K^+.

The lithium-bearing minerals *petalite*, $LiAlSi_4O_{10}$ and spodumene, $LiAl(SiO_3)_2$, are also used as fluxes.

Bone. Strictly speaking, bone functions mainly as a "filler" in bone china bodies and only acts to a limited extent as a flux. It is prepared by treating animal bone with steam to remove fat and calcining at 800°C-1000°C. The product is then wet-ground to reduce 75-80% of the material to less than 10μ and allowed to age for 3-4 weeks, the latter process imparting some degree of plasticity —a useful property in china bodies.

Raw bone is essentially hydroxyapatite, $Ca_5(OH)(PO_4)_3$, with some impurities, notably iron and calcium carbonate; to avoid discoloration of ware, the proportion of iron should be low.

Although the calcium carbonate is decomposed during calcination, there appears to be little decomposition of the hydroxyapatite.

In the firing of bone china, part of the calcium reacts with the whole of the metakaolin, ($Al_2Si_2O_7$), to form anorthite, $CaAl_2Si_2O_8$ (q.v.); the remainder forms tricalcium phosphate, $Ca_3(PO_4)_2$, while the excess phosphate reacts with the fluxes to form a complex glass.

1.7.3 Occurrence of Alkali-bearing Minerals

Deposits of pure feldspar are not very common, being confined principally to Norway, Sweden, the U.S.S.R., and the U.S.A. Rocks containing feldspar in association with other minerals are, however, much more abundant. Examples of such rocks are the granites and pegmatites, which consist mainly of quartz, mica and feldspar cemented together. Unfortunately these rocks are often contaminated with iron-bearing minerals which are undesireable in the manufacture of pottery, since their inclusion discolours the ware.

The chief source of feldspar in Britain is Cornish Stone—a geologically altered granite rock almost free from iron-bearing minerals. Besides felspar, it contains quartz, mica and traces of fluorite, CaF_2; except for the fluorite, none of these impuities adversely affects its value as a flux, since the silica and alumina which they introduce into the body can be allowed for. Fluorite, on the other hand, is troublesome, and no more than about 0.2% should be permitted. It is now removed from Cornish stone by froth flotation. Cornish Stone is extracted by blasting and quarrying. A stone quarry is depicted in Figure 32.

Nepheline syenite is a rock containg the mineral nepheline, $Na(AlSi)O_4$, the structure of which has already been discussed. This rock, which is found only in the U.S.A. and Scandinavia, has usually been purified to some extent before being put on the market. Chemical analyses of some Norwegian nepheline syenites are given in Table 17. Note the high alkali and low iron contents.

Fig. 32.—A stone quarry (photograph by courtesy of English Clays, Lovering Pochin & Co. Ltd.).

1.7.4 Physical and Chemical Properties

Granites and pegmatites are extremely hard grained rocks that are difficult to crush and grind. They are generally light in colour, the shade depending on the chemical composition—e.g. an iron-rich granite is reddish-brown or pink.

Cornish Stone is somewhat softer than granite, depending on the degree of its geological alteration. It contains very little iron oxide, and is therefore an off-white colour; under the microscope, purplish flecks of fluorite can be seen in it. The alteration of granite rock throughout the ages by various agencies results in progressive decomposition of the feldspar to form kaolinite (see later), the quartz and mica remaining unchanged. As might be expected, there are consequently several varieties of Cornish Stone, referred to in order of their degree of alteration as "hard purple," "mild purple," "hard white," and "soft white." The "purple" varieties are richer in feldspar than the "white" varieties and are therefore the more powerful fluxes. In practice, several different varieties may be blended in order to give the body a suitable range of fusion. The chemical compositions of some varieties of Cornish Stone are shown in Table 19. Note the decrease in Na_2O and fluorine (F) from the purple to the white varieties.

Table 17. Analyses of Nepheline Syenite Samples (%)

Sample	Na_2O	K_2O	Fe_2O_3	SiO_2	Al_2O_3	*Loss*
1	6.87	7.07	1.20	52.3	—	2.60
2	7.36	8.80	2.10	53.01	—	1.60
3	7.42	8.20	2.32	53.6	—	1.02
4	7.90	9.08	2.25	52.7	—	1.87
5	8.14	8.75	2.13	52.6	—	1.79
6	7.36	8.80	1.68	53.0	—	1.21

Table 18. Analysis of a Norwegian Feldspar (%)

SiO_2	64.3
Al_2O_3	20.0
Fe_2O_3	0.18
TiO_2	<0.1
CaO	1.5
MgO	0.1
Na_2O	4.9
K_2O	8.4
Li_2O	<0.1
Loss-on-ignition	0.5

Feldspars have no sharp melting-point, but soften between 1140° and 1280°C. Within limits, they lower considerably the softening-point of any aluminosilicate (e.g. a clay) to which they are added, the amount of which depends to a first approximation, on the number of equivalents of Na_2O, K_2O or CaO that have been added. The fluxing-power of alkali-bearing minerals therefore depends on the molecular proportion of Na_2O, CaO or K_2O they contain.

Table 19 Chemical Analyses of Some Varieties of Cornish Stone

Sample No.	*SiO_2*	*Al_2O_3*	*Fe_2O_3*	*TiO_2*	*CaO*	*MgO*	*Na_2O*	*K_2O*	*Li_2O*	*Loss*	*F*
Hard Purple 1	71.9	15.4	0.2	tr.	1.8	0.4	4.1	4.2	tr.	1.4	1.0
2	71.8	15.0	0.2	0.2	1.9	0.2	3.3	4.6		1.6	1.0
3	71.6	15.8	0.1	0.2	2.4	0.1	3.6	3.7		1.7	1.2
4	71.6	15.8	0.2	0.2	2.1	0.1	3.2	4.3		1.9	1.3
Mild Purple 1	72.3	15.4	0.3	0.2	1.7	0.2	3.4	4.1		1.5	0.9
2	72.9	15.5	0.4	0.1	0.8	tr.	0.9	6.9		1.7	0.3
3	72.3	16.1	0.1	0.1	1.8	0.1	5.2	2.1		1.8	0.8
4	71.9	16.0	0.2	0.1	2.1	0.2	3.1	4.1		1.8	1.5
5	72.3	15.4	0.2	0.2	2.1	0.2	2.9	3.9		2.1	1.1
Hard White 1	73.1	16.8	0.2	0.2	0.9		1.2	4.7		2.5	0.3
Soft White	72.8	17.0	0.2	0.2	0.9		2.7	3.8		2.3	0.7

1.8 ALUMINA

1.8.1 Structure

Aluminium oxide or alumina (Al_2O_3) exists in two principal forms— α– and γ–Al_2O_3 (the so-called "β–Al_2O_3" is not the pure oxide but an aluminate having the formula $Na_2O.11Al_2O_3$).

α–Al_2O_3, also known as corundum, is the commonest form of alumina and also the most stable; it is yielded when any other form of alumina is heated to a high enough temperature.

The table of co-ordination numbers (page 7), shows that the radius of the ion Al^{3+} is somewhat greater than that of Si^{4+}, and so it usually has a higher co-ordination number, namely 6. In corundum each Al^{3+} is surrounded by six O^{2-} ions and each O^{2-} ion is surrounded by four Al^{3+} ions, which arrangement achieves electrical neutrality, and the net lattice formula is Al_2O_3. Some idea of the general appearance of the structure can be deduced from Figure 33.

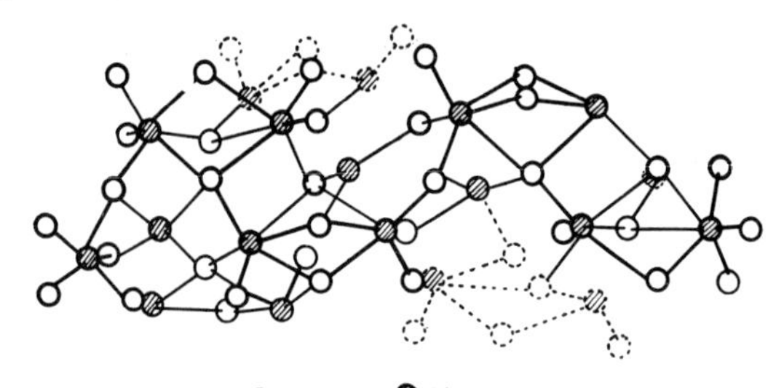

Fig. 33.—The structure of corundum.

γ–Al_2O_3 has a structure based on that of the spinels, which are discussed in the section on chrome, page 98. A spinel has the general formula $A_3B_6O_{12}$, A and B are di- and tri-valent metallic ions respectively. There are nine metallic ions to every twelve O ions in the unit formula. In order to make γ–Al_2O_3 correspond to the spinel structure, the number of O ions must be made up to twelve, as shown, so it must therefore be written Al_8O_{12}. Now in this structure there are clearly only eight metallic ions, compared with nine in the standard spinel formula, which means that out of every nine "places," normally occupied by metallic ions, one is vacant; this is an example of a "defect" lattice. It should be realized, however, that the formula is still electrically balanced, since the eight Al^{3+} ions carry +24 units of charge, and the O^{2-} ions —24 units of charge. Because of these defects in the structure, γ–Al_2O_3 is less stable than α–Al_2O_3 into which it is converted when heated above 1000°C for a long period.

1.8.2 Occurrence

Although aluminium in combinations as alumino-silicate is one of the most abundant constituents of the earth's crust, free alumina is comparatively rare. It is usually found in a hydrated form as bauxite rock, which is actually a mixture of the minerals gibbsite, $Al(OH)_3$, diaspore and boehmite, (both of formula $HAlO_2$). Bauxite is found in Jamaica, British Guiana, Europe and Russia and elsewhere. Non-hydrated alumina, i.e. corundum, is a rare mineral, found in Greece and South Africa. Bauxite rock has been formed by the very severe weathering of various kinds of igneous rocks under tropical conditions, in which circumstances not only the alkali oxides but also the silica was removed, leaving only hydrated oxides of aluminium and iron; this residue formed bauxite rock. The bauxite clays of Ayrshire are said to have originated in a similar way.

Commercial alumina is produced chiefly from bauxite, the total world production being millions of tons per year. The bauxite ore contains, besides oxides of aluminium, silica and ferric oxide, which are removed by the *Bayer method* as follows. First the ore is ground fine, and it is then treated with sodium hydroxide solution in an iron autoclave under a pressure of 4 atm and at 160°—170°C. The alumina dissolves, forming sodium aluminate:

$$Al_2O_3 + 2NaOH = \underset{\text{Sodium aluminate}}{2NaAlO_2} + H_2O$$

The silica dissolves to form sodium silicate, but the ferric oxide remains undissolved, as the so-called "red mud," which is filtered off thereby freeing the solution from iron.

The sodium aluminate is unstable and is readily decomposed by passing carbon dioxide gas through the solution:

$$2NaAlO_2 + CO_2 + 3H_2O = Na_2CO_3 + \downarrow 2Al(OH)_3$$

The final product, aluminium hydroxide, $Al(OH)_3$, is separated by filtration and washed. The sodium silicate remains in solution and so is removed. The aluminium hydroxide is finally calcined at 1000°C or higher, when it loses water of constitution, yielding alumina :

$$2Al(OH)_3 \xrightarrow{1000C^\circ} Al_2O_3 + 3H_2O$$

It is desirable to calcine sufficiently so that α–Al_2O_3 and not γ–Al_2O_3 is formed. Since bauxite ores are comparatively rare, efforts are being made to extract alumina from clays. High-grade commercial alumina still contains 0.1—0.2% of Na_2O, 0.1% of CaO and traces of Fe_2O_3, TiO_2 and Cr_2O_3.

1.8.3 Physical Properties

Natural corundum or α–Al_2O_3 is a water-white, very hard, crystalline substance. Gem stones such as sapphire and ruby consist chiefly of corundum, with traces of other oxides.

Commercial alumina is a white powder, with a specific gravity about 3.9 which, if it has been calcined sufficiently, consists of minute crystals of α–Al_2O_3. It can be shaped by pressing or slip-casting and when the shapes are fired about 1700°—1800°C they become remarkably strong, even though the firing-temperature is well below the melting-point (2050°C). This process is known as *sintering* and is employed in the manufacture of various pure-oxide ceramics.

Articles of sintered alumina are hard, not readily attacked by acids or alkalis at high temperatures, and can withstand considerable changes of temperature without fracturing, consequently sintered alumina is used for making crucibles, thermocouple sheaths and sparking-plugs; it also has a very high electrical resistance and so insulators are made from it.

Raw bauxite is frequently added to fireclay to increase its alumina content; the high-alumina bricks made from such mixtures have a high resistance to fusion.

1.8.4 Chemical Properties

As already mentioned, α–Al_2O_3 is very inert and resists most aqueous acids and alkalis. Fused caustic alkalis attack alumina slowly, forming aluminates. In analysing alumina, in order to ensure complete reaction it is necessary to fuse with borax or with sodium peroxide.

γ–Al_2O_3 is rather more reactive than α–Al_2O_3, and can be dissolved by hot concentrated acids. Articles are often made from α-Al_2O_3 by slip-casting, using deflocculated aqueous suspensions. The electrical charge on the surface of an α-Al_2O_3 particle is determined by the manner in which the amphoteric surface groups ionize. In aqueous suspension, the surface $\gtrdot$Al-O- groups are hydrated to —Al-O-H, which in acid solution release OH^- ions, replaceable by Cl^- or other anions:—

$$\gtrdot Al\text{-}O\text{-}H \longrightarrow \gtrdot Al^+ + OH^-$$

leaving a resultant positive charge on the surface. However, in alkaline solution, H^+ ions are released, replaceable by other cations, and a negative charge is formed on the surface:—

$$\gtrdot Al\text{-}O\text{-}H \longrightarrow \gtrdot Al\text{-}O^- + H^+$$

Accordingly, Al_2O_3 suspensions may be deflocculated by HCl (at about pH 4), when Cl^- replaces OH^- as the counter-ion; or by NaOH and other bases, when the appropriate cation replaces H^+.

Most other oxides, being more basic than Al_2O_3, ionize in water to produce positively-charged particles and are therefore generally deflocculated by HCl. In some instances, better casts are obtained with suspensions in alcohol or other organic media, with fatty acids as deflocculants.

1.9 OTHER REFRACTORY RAW MATERIALS

1.9.1 Magnesite

Magnesite is one of the group of basic refractories that were introduced in 1880 to cope with problems encountered in steelmaking. The term "magnesite" really means magnesium carbonate, $MgCO_3$, which occurs in sedimentary rocks formed by the decomposition of ultrabasic rocks containing olivine and other magnesium silicates. It is often associated with limestones and dolomites, and is found in Austria, Czechoslovakia, Greece, Yugoslavia, Russia, Canada the U.S.A., Brazil and Manchuria. There are only a very few minor deposits of magnesite in the British Isles, but sea-water magnesite has superseded natural magnesite in the British ceramic industries.

Production of Sea-water Magnesite

Sea-water magnesite is unique in that it is obtained by a chemical reaction, the basis of which is as follows. Sea-water contains, among other things, sulphates and chlorides of sodium, potassium, calcium and magnesium. If calcium hydroxide is added to the sea-water, the magnesium sulphate and chloride react to form sparingly-soluble magnesium hydroxide, which is precipitated, and calcium chloride and sulphate as by-products. The other salts are unaffected and remain in solution. The calcium hydroxide required for the process is obtained by calcining dolomite, $CaMg(CO_3)_2$, to give CaO and MgO, and slaking the latter with water to form the hydroxides :

$$\underset{\text{Dolomite}}{CaMg(CO_3)_2} \xrightarrow{\text{heated}} CaO + MgO \xrightarrow{+\ H_2O} \underset{\text{“Slaked dolomite”}}{Ca(OH)_2 + Mg(OH)_2}$$

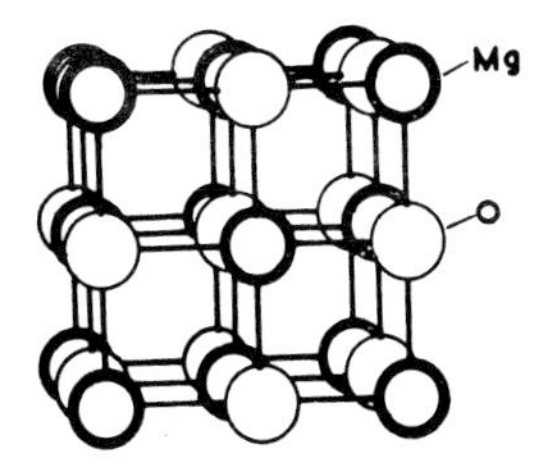

Fig. 35.—The structure of magnesium oxide.

Fig. 34.—Production of sea-water magnesite at Hartlepool : one of the settling tanks, (photograph by courtesy of the Steetley Magnesite Co. Ltd.).

This "slaked dolomite" is then allowed to react with sea water, in reaction tanks, which are continuously agitated. The precipitated magnesium hydroxide is then concentrated by settling, to form a thick sludge, dried, and calcined at 1600°C (Figure 34).

The reaction may be represented by the equation :

$$\underbrace{\begin{bmatrix} Mg(OH)_2 \\ Ca(OH)_2 \end{bmatrix}}_{\text{"Slaked dolomite"}} + \underbrace{\begin{bmatrix} MgCl_2 \\ MgSO_4 \end{bmatrix}}_{\text{Sea-water}} \longrightarrow \underbrace{2Mg(OH)_2}_{\text{Magnesium hydroxide}} + \underbrace{\begin{bmatrix} CaCl_2 \\ CaSO_4 \end{bmatrix}}_{\text{By-products}}$$

Preparation of Natural Magnesite

Before use, natural magnesite is always calcined decomposing to form magnesium oxide and CO_2 :

$$MgCO_3 \xrightarrow{\text{calcined at } 1500°\text{—}1600°\text{C}} MgO + CO_2$$

This magnesium oxide is used for making refractories, but despite the chemical change during calcination, it is still referred to as "magnesite." Care must be taken, therefore, to avoid confusing "natural magnesite" with "calcined magnesite."

Just as quicklime, CaO, slakes or combines with water, MgO

tends, in the presence of atmospheric moisture, to combine with water, forming the hydroxide, $Mg(OH)_2$. This tendency is considerably reduced, however, by calcining at 1500°C or over, when it is said to be dead-burned.

Structure of Magnesium Oxide

The crystal structure of magnesium oxide is illustrated in Figure 35. This is one of the simplest structures and is very similar to that of rocksalt, NaCl. Each Mg atom is surrounded by six O atoms, and to preserve electrical neutrality, each O atom is surrounded in turn by six Mg atoms. The entire lattice can be considered as built up from cubes, with O and Mg atoms at alternate corners; this is the only known crystalline form of MgO and is sometimes called *periclase*.

Physical Properties

Pure magnesium oxide is a white powder, having a specific gravity of 3.65 when calcined, and a melting point of 2800°C.

The refractoriness of commercial magnesite refractories is considerably lower than that of the pure oxide, owing to the presence of impurities, which in fact act as a bond in the manufacture of magnesite bricks. Magnesite is used extensively as a lining for the hearth in basic open-hearth furnaces.

Chemical Properties

The chief value of calcined magnesite lies in its resistance to attack by basic slags containing CaO and Fe_2O_3. On the other hand, it is susceptible to attack by acidic oxides such as SiO_2, combining with them to form low-melting compounds.

Even after it has been calcined, magnesite is attacked by strong acids; this property is made use of in the analysis of magnesite.

1.9.2 Dolomite

The mineral dolomite is the double carbonate of calcium and magnesium, having the formula $CaMg(CO_3)_2$. Over a million tons of the raw material are processed annually for the construction and maintenance of the linings of steel-melting furnaces. Its replacement by magnesite, the chemical resistance of which is higher, has so far progressed to only a small extent.

The preparation of dolomite for use as a refractory is similar to that for magnesite—the raw material is calcined, a mixture of calcium oxide and magnesium oxide being formed, with the liberation of CO_2.

Dolomite or Magnesian Limestone

Dolomite occurs in the Permian Magnesian Limestone and the Carboniferous Limestone (Table 9). The Permian deposit runs along the eastern side of the Pennines from Nottingham in the south to Coxhoe and Hartlepool in the north. Deposits of dolomite are found also in South Wales.

Dolomite rocks have been formed by a number of complex reactions, two of which may be mentioned: (1) the action of magnesium-containing solutions on limestone, (2) the precipitation of magnesium carbonate from super-saturated solutions.

Structure

The structure of calcined dolomite is essentially that of periclase and calcium oxide, both of which have the rocksalt structure (Figure 35).

Physical Properties

The calcined product is normally available in porous granular form of bulk density 2.6—3.0 g/ml.

Chemical Properties

The principal disadvantage of calcined dolomite is its tendency to hydrate, causing bricks made from it to crumble. This hydration tendency is due to the high proportion of lime and persists even after having been calcined to 1600°C. To overcome this, dolomite can be stabilized by being fired with a silicate such as steatite (talc) which combines with the free lime to form calcium silicate; partial stabilization can be achieved by water-proofing the bricks with tar.

Dolomite is slightly more susceptible to attack by iron-rich slags than is magnesite, but otherwise it is very similar to magnesite in its chemical properties.

1.9.3 Chrome

Chrome ore is a basic rock containing chromium, iron, aluminium, magnesium and oxygen combined together as a complex crystalline compound called a *spinel*. In addition, the mineral serpentine, $Mg_3Si_2O_5(OH)_4$, is present as an impurity, sometimes called the "gangue."

Structure

Spinels have the general formula AB_2O_4, where A stands for a divalent ion (e.g. Fe^{2+}, Mg^{2+}, Cu^{2+}, Mn^{2+}) and B for a trivalent ion (e.g. Cr^{3+}, Fe^{3+}, Al^{3+}). Chromite, $FeCr_2O_4$, magnetite, $Fe^{2+}\ Fe_2^{3+}\ O_4$, and spinel itself, $MgAl_2O_4$, all belong to this group.

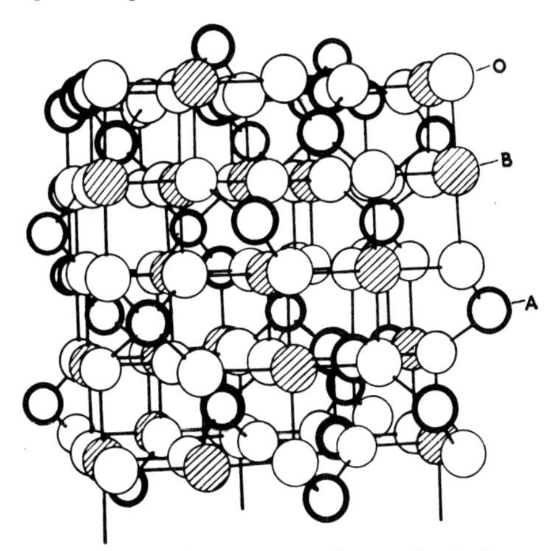

Fig. 36.—The structure of normal spinel.

In a *normal* spinel, the A, B and oxygen atoms are so packed that each A atom is surrounded by four O atoms, and each B atom by six O atoms (Figure 36).

There is, however, another type of spinel, called an *inverse* spinel, in which the positions normally occupied by A atoms are taken by half the B atoms, the A atoms then occupying the original "B positions." If we denote the "oxygen co-ordination" of each atom by a Roman numeral thus :— A^{IV}, then a normal spinel can be written $A^{IV}B_2{}^{VI}O_4$ and an inversed spinel $B^{IV}(AB)^{VI}O_4$.

Ferrous chromite, $Fe^{IV}Cr_2{}^{VI}O_4$, is a normal spinel; magnetite, $Fe^{IV}(Fe_2{}^{VI}O_4)$ is an inversed spinel, one of the 6-co-ordinated Fe atoms being divalent, the other trivalent. In chrome ore, the principal constituent is a mixed spinel in which the "A"-positions are occupied by Fe^{2+} and Mg^{2+}, and the "B"-positions by Cr^{3+}, Al^{3+} and Fe^{3+}.

Unlike magnesite, chrome ore retains essentially the same structure (spinel) when heated, although probably some of the ferrous iron is oxidised; at the same time, the siliceous impurities are decomposed and afterwards recombine to form new products.

Occurrence and Uses of Chrome Ore

Chrome ore is found in Russia, the Phillippines, North America, Rhodesia, Turkey and Greece, and is usually extracted by quarrying.

Chrome refractories were first introduced in about 1880 as a neutral zone between the acid and basic courses of steel-making furnaces. Nowadays, chrome ore is usually mixed with dead-burned magnesite for the production of *chrome-magnesite*, a very important basic refractory, which has made the "all-basic" open-hearth steel furnace an economic possibility.

Physical Properties

Chrome ore is a reddish-brown mineral, having a specific gravity (depending on its composition) about 4.3—4.6. Being a mixture, it has no well-defined melting-point, but is said to soften over about 1700°C. Under a load of 50 lb./in^2, however, it fails completely between 1370° and 1470°C.

Chemical Properties

Chrome ore is a comparatively inert substance. It is but little attacked by basic oxides or acidic oxides at high temperatures and for this reason it is used as a neutral zone in furnaces containing both acid and basic refractories.

The ore can be dissolved by the action of hot, concentrated sulphuric acid, which property is made use of in the analysis of chrome ores. It is, however, only incompletely dissolved by the other common acids, and resists the action of caustic alkalis.

1.10 MISCELLANEOUS MATERIALS

1.10.1 Rutile or Titanium Dioxide, TiO_2

Rutile is a mineral associated with sands and sandstones; it is found in Travancore, India, and Australia. It is relatively inert, has a specific gravity of 4.2, is insoluble in water and acids, except sulphuric acid but is soluble in caustic alkalis.

The chief interest in rutile lies in its high dielectric constant (about 110), which makes it a valuable material for electrical condensers. With other oxides it combines to form *titanates*, which have even higher dielectric constants (several thousand).

1.10.2 Zirconia or Zirconium Dioxide, ZrO_2

Zirconia occurs as the minerals zircite and baddeleyite, in Brazil and Ceylon. Pure zirconia has a specific gravity of 6.3, and is insoluble in water but soluble in sulphuric acid and hydrofluoric acid. It has a very high melting-point (about 2700°C) and is therefore used as a refractory material, although its cost prevents its being used on a large scale. Like silica, zirconia undergoes an inversion, accompanied by a volume change at 1000°C., but may be stabilized by the addition of 10—20% of CaO.

Zirconium silicate or zircon, $ZrSiO_4$, found in India, Madagascar and Australia, is used in the manufacture of zircon porcelain for electrical insulation.

1.10.3 Beryllia or Beryllium Oxide, BeO

The chief source of beryllium oxide is the mineral beryl, or beryllium aluminium silicate, $3BeO.Al_2O_3.6SiO_2$, which occurs in Argentina, Brazil and India. When pure, beryllium oxide has a specific gravity of 3.0, and is almost insoluble in water; it is soluble in sulphuric acid and in fused alkalis. It has been used as a refractory (melting point over 2500°C) but its toxicity has restricted its use.

READING LIST

"Structural Inorganic Chemistry", A. F. WELLS, (Oxford, 1950; recent edition, 1961).
Chapter on silica and silicates.

"Introduction to Physical Chemistry", A. FINDLAY, (Longmans, 1941).
Chapter 11 : "The Colloidal State".

"Steelplant Refractories", J. H. CHESTERS, (The United Steel Companies, Ltd., Sheffield, 1963).
Chapters 3, 4 and 5, relating to the preparation of raw materials dolomite, magnesite and chrome.

"Heavy Clay Technology", F. H. CLEWS, (British Ceramic Research Association, Stoke-on-Trent, 1955).
Chapters on extraction of clay and physical properties of clay.

"Elements of Ceramics", F. H. NORTON, (McGraw-Hill Book Co. Inc., New York, 1949).
Chapters on structure of matter and extraction of raw materials.

ACKNOWLEDGMENTS

The author acknowledges the valuable assistance given to him by the textbook sub-committee of the Institute of Ceramics.

He is also indebted to the British Ceramic Society and the British Ceramic Research Association for much useful information.